WATER LAW, POVERTY, AND DEVELOPMENT

Water Law, Poverty, and Development

Water Sector Reforms in India

By
PHILIPPE CULLET

UNIVERSITY PRESS

OXFORD
UNIVERSITY PRESS

Great Clarendon Street, Oxford OX2 6DP

Oxford University Press is a department of the University of Oxford.
It furthers the University's objective of excellence in research, scholarship,
and education by publishing worldwide in

Oxford New York

Auckland Cape Town Dar es Salaam Hong Kong Karachi
Kuala Lumpur Madrid Melbourne Mexico City Nairobi
New Delhi Shanghai Taipei Toronto

With offices in

Argentina Austria Brazil Chile Czech Republic France Greece
Guatemala Hungary Italy Japan Poland Portugal Singapore
South Korea Switzerland Thailand Turkey Ukraine Vietnam

Published in the United States
by Oxford University Press Inc., New York

First published 2009

British Library Cataloguing in Publication Data
Data available

Library of Congress Cataloging in Publication Data
Cullet, Philippe.
Water law, poverty, and development: water sector reforms in India / by Philippe Cullet.
p. cm.
Includes bibliographical references and index.
ISBN 978-0-19-954623-7
1. Water—Law and legislation—India. 2. Water-supply—Government policy—India.
3. India—Economic policy. I. Title
KNS2522.C85 2009
346.5404′691—dc22 2009016861

Typeset by Newgen Imaging Systems (P) Ltd., Chennai, India
Printed in Great Britain
on acid-free paper by the
MPG Books Group, Bodmin and King's Lynn

ISBN 978–0–19–954623–7

1 3 5 7 9 10 8 6 4 2

Summary Contents

Contents

Abbreviations and Hindi Terms

ADB	Asian Development Bank
ARWSP	Accelerated Rural Water Supply Programme
CPHEEO	Central Public Health and Environmental Engineering Organization
dalit	term describing people formerly called 'untouchables' or 'outcastes'
Delhi Jal Board	Delhi Water Board (or DJB)
FAO	Food and Agriculture Organization of the United Nations
GATS	General Agreement on Trade in Services
GATT	General Agreement on Tariffs and Trade
gram panchayat	village council
gram sabha	a body consisting of persons registered in the electoral rolls relating to a village comprised within the area of a panchayat at the village level (Article 243, Constitution)
ICESCR	International Covenant on Economic, Social and Cultural Rights
IWRM	integrated water resources management
jal sansthan	water board
JNNURM	Jawaharlal Nehru National Urban Renewal Mission
johad	traditional rainwater storage tank
lpcd	litres per capita per day
MLA	Member of the Legislative Assembly (state legislature)
panchayat	institution of self-government constituted under Part IX of the Constitution
panchayati raj institutions	institutions of self-government for rural areas, whether at the level of a village or of a block or district
PMU	Project Management Unit
RGNDWM	Rajiv Gandhi National Drinking Water Mission

Rs	Rupees
sarpanch	chairperson of a panchayat (Rajasthan)
SC	scheduled castes
SRP	Sector Reform Project
ST	scheduled tribes
UIDSSMT	Urban Infrastructure Development Scheme for Small and Medium Towns
UNCED	United Nations Conference on Environment and Development
UNECE	United Nations Economic Commission for Europe
UNECE Water Convention	Convention on the Protection and Use of Transboundary Watercourses and International Lakes 1992
UNEP	United Nations Environment Programme
UN Water Convention	Convention on the Law of the Non-navigational Uses of International Watercourses 1997
Uttaranchal	former name for the state of Uttarakhand
VWSC	village water and sanitation committee
WMO	World Meteorological Organization
WSSD	World Summit on Sustainable Development
WUA	water user association

Hindi words referred to in the text, whether commonly used in English or not, have not been italicized.

Table of Cases

Introduction

Freshwater is fundamental for life on earth and for human life specifically. Thus, access to sufficient water of appropriate quality is one of our most immediate and basic requirements. Its fulfilment is required on a daily basis as a matter of simple survival. This explains why water has always been at the centre of policy concerns for most societies throughout human history. In fact, water availability has been a precondition for the development of human settlements and natural or human-induced water stress has played an immense role in their decline.

The importance of water as a basic survival need, as a livelihood, as an input for agriculture and hence food sovereignty and as an input for economic development related activities have given it a central role in governance for a long time. Thus, whether it is aqueducts built during the Roman Empire, irrigation tanks in South India or the large dams of the twentieth and twenty-first century, the harnessing of water for drinking or other activities has been of central importance. The essential nature of water for human survival led most societies to give it a special status in law. One of the primary consequences of this special nature was the introduction of restrictions on the ownership over water per se.

In keeping with the central role of water, its regulation has been the object of significant attention for a long time. During colonial times, significant emphasis was put on the development of irrigation law. This has had lasting impacts because some of the basic principles deemed to govern water overall are only found in irrigation acts from the colonial period or acts inspired by the same ethos. Yet, in India, water law is comparatively underdeveloped. Thus, to-date there is no framework water legislation. Similarly, there is no drinking water legislation even though this is, in principle, a primary concern of the Union and all state governments.

Besides formal water legislation, a number of other water-related principles have emerged over time. These include, for instance, principles concerning the rights of landowners on neighbouring flowing water and groundwater. More recently, water pollution and more generally the environmental impacts of water use have been the object of increasing attention. While different uses of water have been separately regulated, there has been no comprehensive water legislation addressing all needs and uses. This did not prove problematic as long as human-induced and natural water stress was not too prevalent. This is not the case any more and the existing legal framework is largely incapable of addressing the new challenges that have arisen over the past few decades.

Water law in India has been heavily influenced by early developments of water legislation that focused on the economic potential of water and the types of rights that landowners could claim over water found underneath or adjoining their plots. It thus remains straight jacketed in a conceptual framework, which is, by definition, incapable of taking on board the multiplicity of water uses and, in particular, the fact that it is a human right. A broader framework needs to be provided for water law. The premise must be that water is first of all essential for life, essential for the realization of a number of our most fundamental human rights, essential for all ecosystems and extremely important as a resource for a variety of activities ranging from food production to energy generation. This gives water law a prime role in poverty eradication and the realisation of a socially equitable and environmentally sustainable process of development. In other words, water law must be at the centre of any broadly conceived strategy of development.

The need for a broader framework is, in principle, a relatively innocuous assertion in an era that has seen international and national law move rapidly over the past decades to effectively include human rights, environmental conservation, equity and sustainability dimensions. Yet, it raises a number of questions in the context of water. Firstly, while the international law of sustainable development has rapidly developed over the past two decades, water is remarkably absent from the areas that have received most attention. In fact, international water law remains dramatically underdeveloped since states have neither moved beyond the traditional focus on international watercourses nor even managed to ratify the rather conservative convention adopted in 1997 after years of negotiations. Thus, a country like India cannot expect to get much effective guidance from international water law in the development of a modern national water law that addresses all the dimensions of water. Secondly, water, like biodiversity, can neither be comprehensively addressed only at the national level nor only at the international level. In this sense, it is typical of a new emerging pattern of transnational law which needs to be conceived, implemented and analysed from the local to the global level as if it were one single framework to ensure its coherence and effectiveness. Thirdly, while international water law is of comparatively little importance in ongoing water law reforms, these are strongly influenced by a policy consensus around water issues. A series of non-binding instruments not necessarily developed in UN frameworks or legislatures thus yields a disproportionate influence on ongoing changes in water law.

The need for changes to Indian water law stems from a variety of reasons. These include the narrow framework within which existing water law is conceived, outdated principles that do not reflect existing scientific knowledge, for instance, concerning groundwater and the new challenges that have surfaced over the past few decades, such as the dramatic rise in groundwater consumption triggered by the introduction of mechanical pumping devices. All these are compelling reasons for introducing water law reforms. Yet, in practice, no broad-based movement towards water law reform has taken place. Rather, the process of water law reform that is ongoing has been conceived as part of what are known as water sector reforms and actual reforms introduced have been sectoral.

Water sector reforms are a set of measures proposed to ensure that the water sector can address the challenges of the twenty-first century. Water sector reforms are at the same time broadly conceived and based on a narrow understanding of the problems that need to be addressed. On the one hand, they suggest that water should be managed in an integrated manner, which includes, for instance, the need to address water issues at the basin level. On the other hand, water sector reforms tend to focus on issues of governance and management. Thus, even though water sector reforms suggest that the reform process aims at maximizing economic and social welfare, the real emphasis is on economic aspects. This includes, for instance, the call for turning water into an economic good, a fundamental change that will have immense repercussions for many years. Water sector reforms also specifically advocate private sector participation and the reduction of the role of the state in the water sector. On the whole, adopting water sector reforms implies significant changes because the principles proposed are different from existing principles governing the water sector.

Water sector reforms have been implemented in a variety of ways over the past couple of decades through specific projects or schemes. Little emphasis was put on the law even though certain proposals, like the introduction of water rights, have significant legal implications. This can be partly explained by the fact that some of the reforms could proceed without amending existing legal frameworks or without introducing new laws. Yet, this model was found to be partly unsuccessful, for instance, because of failed privatization schemes and partly lacking because it did not sufficiently entrench reforms beyond the specific contexts where they were introduced.[1] The perceived need to consolidate project-specific reforms and to introduce specific legal frameworks that incorporate the principles of water sector reforms was taken up as a way to ensure the permanence of reforms. This new strategy is particularly visible in India where the introduction of water law reforms has been taken up as a way to strengthen processes already started and to initiate new changes. As a result, water law reforms have emerged as a central component of the overall process of reform in the past decade. One of the peculiarities of these new laws is that their adoption is often linked to the implementation of a project spearheaded either by the World Bank or the Asian Development Bank. In certain cases, some of the conditions of the loan specifically include the adoption of certain laws while in other cases legal changes may be part of a broader set of

[1] eg World Bank, Efficient, Sustainable Service for All? An OED Review of the World Bank's Assistance to Water Supply and Sanitation (Report No. 26443, 2003) 21.

measures, such as in Maharashtra where two major new acts were adopted just before the signing of the agreement for a big World Bank water project.

The new focus on law reform in the context of water sector reforms has two main consequences for water law. Firstly, the pace of change has dramatically increased over the past decade with the swift adoption of a number of new laws. Secondly, water law has become a central component of the broader set of reforms in the water sector. Both factors indicate that water law should be given significant attention by all concerned actors, from all individuals concerned with the implementation of the human right to water to panchayats and municipalities, civil society organizations, state governments, state and national policy makers, academics, and national level institutions. In reality, while certain issues like the reform of patent law in the wake of the adoption of the Agreement on Trade-Related Aspects of Intellectual Property Rights have been the object of significant public debate throughout society for the past decade,[2] the same thing has not happened with water yet. This is unexpected given that the realization of the human right to water is something even more immediate than concerns over the right to health that surfaced in the context of the introduction of product patents in the health sector.[3]

The lack of broad public debate over water law reforms, the lack of academic literature and the limited involvement of civil society organizations in water law issues are even more surprising given the intense interest that water in general elicits. Thus, water sector reforms have been the object of numerous studies and books over the past couple of decades.[4] Water law, however, has not been the object of much interest since the early 1990s and most of the existing literature focuses on inter-state aspects.[5] The general lack of interest for water law has meant that few people were ready to pick up and analyse the developments ushered by the introduction of water law reforms. This is unfortunate because the reforms that are being introduced are momentous and will redefine water law in India for many years to come.

[2] eg R Dhavan & M Prabhu, 'Patent Monopolies and Free Trade: Basic Contradiction in Dunkel Draft' (1995) 37 *J Indian L Institute* 194 and NS Gopalakrishnan, 'Protection of Traditional Knowledge—The Need for a *Sui Generis* Law in India' (2002) 5 *J World Intellectual Property* 725.

[3] On the relationship between patents and health, S Chaudhuri, *The WTO and India's Pharmaceuticals Industry: Patent Protection, TRIPS and Developing Countries* (New Delhi: Oxford University Press, 2005).

[4] eg A Gulati, R Meinzen-Dick & KV Raju, *Institutional Reforms in Indian Irrigation* (New Delhi: Sage, 2005), GN Kathpalia & R Kapoor, Water Policy and Action Plan for India 2020: An Alternative (Delhi: Alternative Futures, 2002), V Pangare, N Kulkarni & G Pangare, An Assessment of Water Sector Reforms in the Indian Context: The Case of the State of Maharashtra (Geneva: UNRISD, 2004), K Prasad (ed), *Water Resources and Sustainable Development—Challenges for the 21st Century* (New Delhi: Shipra, 2003) and M-H Zérah, *Water—Unreliable Supply in Delhi* (New Delhi: Manohar, 2000).

[5] The landmark study of the early 1990s was C Singh, *Water Rights and Principles of Water Resources Management* (Bombay: Tripathi, 1991). On inter-state and international aspects, eg R D'Souza, *Interstate Disputes over Krishna Waters—Law, Science and Imperialism* (New Delhi: Orient Longman, 2006) and SP Subedi (ed), *International Watercourses Law for the 21st Century—The Case of the River Ganges Basin* (Aldershot: Ashgate, 2005).

This book focuses on India for several reasons. Firstly, the widespread process of water law reform that is currently under way requires much more attention than it has been given until now. Secondly, India is one of the first countries where so much emphasis has been put on law reform in the context of water sector reforms. Thirdly, the fact that water is a state subject means that most reforms are taken up at the state level. Given the diversity of hydrological, environmental and socio-economic conditions that mark the different states, the Indian experience is in principle full of lessons for other countries. Fourthly, India is not only one of the biggest countries in the South but also a country that is a microcosm of the broader world. While it has experienced fast economic growth for more than a decade, this growth has not trickled down and the overwhelming majority of the population still experiences severe poverty, as illustrated by the fact that 77 per cent of the population lives with less than Rs 20 a day.[6] It thus exhibits traits of a rapidly developing economy as well as traits of a poverty-ridden country at the same time.

This book seeks to contribute to a better understanding of ongoing reforms, their rationale, their broader context and their likely impacts. The rapidity of the changes that are unfolding together with the lack of an ongoing body of scholarship focused specifically on water law reforms have not yet led to the development of sufficient literature on these new issues.[7] In this context, the main question that arises is not whether water law reforms are necessary but what kinds of reforms should be introduced. There has, however, been no broad debate on the type of reforms that India needs. This work thus seeks to contribute to the critical analysis of ongoing water law reforms so as to ensure that the framework that informs the reforms and the consequences of the decisions taken are clear to all concerned. Additionally, it shows that there exist different options for water law reform. It is imperative that all these options be further examined, debated and critically analysed before more long-term decisions are taken.

This work examines the reforms that are taking place in India. Yet, by the very nature of the reforms being introduced, the analysis builds to a significant extent on the international policy framework that is influencing developments at the national level. This is, however, not a traditional study examining the ways in which international law is taken into account in the formulation of domestic law. In fact, existing international water law only addresses the issues that are at the centre of this study tangentially. The frameworks that are

[6] National Commission for Enterprises in the Unorganised Sector, Report on Conditions of Work and Promotion of Livelihoods in the Unorganised Sector (New Delhi: Ministry of Small Scale Industries, Government of India, 2007).

[7] Renewed interest in water law is highlighted, for instance, by the publication of R Iyer (ed), *Water and the Laws in India* (New Delhi: Sage, forthcoming 2009) and P Cullet, A Gowlland-Gualtieri, R Madhav & U Ramanathan (eds), *Water Law for the Twenty-first Century: National and International Aspects of Water Law Reforms in India* (Abingdon: Routledge, forthcoming 2009).

most relevant are either soft law instruments or the strategies and policies of development banks. The interactions between the national and the international levels are thus much more complex than the traditional framework of interaction between national and international law.

The analysis of ongoing reforms indicates that the conceptual framework that informs ongoing water law reforms is not sufficiently broad-based to reflect the multiplicity of water uses and its different functions, in particular its human right, social and environmental dimensions. This is, for instance, borne out by the analysis of law and policy reforms for drinking water that adopt the same economic model proposed for other water uses and fails to focus on the realisation of the human right to water for the poorest and most marginalised despite the general poverty eradication framework of water sector reforms. This points to the need for proposing alternative bases for water law reforms. Since water law reforms of one kind or another are imperative at this juncture, this book proposes a series of alternative bases for broad-based reforms starting from a human rights, equity, and sustainability perspective.

Proposed changes in water law suggested in this book focus on the national level. Similar proposals could be made at the international level where water law also needs to be further developed. Yet, at this juncture, it does not seem realistic to expect the prompt adoption of international water treaties focusing primarily on the social, human rights, and environmental aspects of water. It is rather incremental changes at the national level in various countries that may foster the progressive development of a body of water law that could constitute the basis for later developments at the international level.

The structure and organization of this book is influenced by the context within which it has been conceived. The relative lack of recent scholarship on water law and the absence of a body of work concerning ongoing water law reforms implies that a sizable part of the sources used are primary documents and literature in other disciplines. Additionally, since a number of the most important reforms are recent or unfolding, there is often little that can be said yet concerning their actual implementation. As a result, part of the analysis is limited to an analysis of the legal frameworks that have been adopted and the conceptual framework that informs them. Since a broader analysis of the changes that are taking place is needed, the impacts of policy changes in the specific context of drinking water are analysed with regard to policy frameworks as well as their implementation in a few selected places. Further work will be required in years to come to analyse, for instance, the actual impacts of the introduction of independent regulators in practice, something which cannot be undertaken yet.

Chapter 1 starts by providing a general background on the water situation, water uses and the policy background that informs water law. Chapter 2 moves on to examine water law as it has developed until the introduction of the current water law reforms. Chapter 3 then examines water sector reforms and, in particular, the conceptual framework that informs them. Chapter 4 analyses some of the main law reforms currently proposed and implemented in different states in India. Chapter 5 carries on the enquiry by focusing on the specific situation of drinking water. Following the critical analysis of ongoing water law reforms, chapter 6 suggests a series of alternative bases for water law reforms. These are necessary to ensure that water law focuses primarily on the human right and social dimensions of water rather than its economic aspects.

1

Context for Water Law and Water Sector Reforms

Water is the substance that makes life on Earth possible. It is so important that it gives the planet its nickname. Yet, the overall abundance of water masks the fact that all terrestrial ecosystems rely exclusively on freshwater which constitutes only 2.5 per cent of all water on the planet.[1] Most of this is held in the form of permanent ice or snow, with Antarctica and Greenland accounting for 68 per cent of all freshwater.[2] This leaves only 1 per cent of all freshwater—or 0.01 per cent of all water—as water that is directly available for human consumption.[3] This water resides largely in lakes, rivers, wetlands and shallow groundwater aquifers.[4] Groundwater constitutes 96 per cent of unfrozen freshwater.[5] This amount of water is regularly renewed by rainfall and snowfall. Most people and most countries are thus highly dependent on the global water cycle for most of the water they use on a regular basis.[6]

Freshwater is currently available in sufficient quantities at a global level. However, even though water withdrawals amount to only 10 per cent of global freshwater runoff, this is equivalent to 54 per cent of readily accessible water.[7] Further, it is the tripling of demand for freshwater over the past 50 years that has already pushed water withdrawals beyond the recharge potential of many aquifers.[8] In India, the total actual water resources availability is estimated at

[1] United Nations, *Water—A Shared Responsibility* (Paris: UNESCO, 2006) 121.

[2] United Nations, *Water for People—Water for Life* (Paris: UNESCO, 2003) 68.

[3] United Nations Environment Programme, *Global Environment Outlook 3* (London: Earthscan, 2002) 150.

[4] CJ Vörösmarty, 'Fresh Water' in R Hassan, R Scholes & N Ash, *Ecosystems and Human Well-being: Current State and Trends—Millennium Ecosystem Assessment Series Volume 1* (Washington: Island Press, 2005) 170.

[5] United Nations (n 1 above) 128.

[6] For a discussion of the consequences in law, see ch 6.B.1.

[7] WWF, Living Planet Report (Gland: WWF, 2006) 12.

[8] L Brown, *Plan B 2.0—Rescuing a Planet Under Stress and a Civilization in Trouble* (New York, NY: WW Norton, 2006).

around 1,869 billion cubic metres.[9] At present 34 per cent of this amount is used.[10] With regard to groundwater, despite a tremendous increase in the number of wells from 3.8 million in 1951 to 17.3 million in 1997, the overall state of groundwater development is 58 per cent at present.[11]

Availability of water is an issue which is eminently local because it remains relatively difficult to transport significant amounts of water over long distances. It is also clearly national, regional and global in scope. This is, for instance, borne out by the fact that from ancient times until today, water has been transported over long distances to supply water to cities. This is as true of Roman aqueducts transporting water over long distances as of the Tehri dam in the Himalayas supplying water to the city of Delhi. Further, water availability in specific localities depends on river flows coming from other regions. More generally, water availability is linked to the broader water cycle, which is itself heavily dependent on climatic conditions. Thus, in many parts of the world, precipitations are the main direct and indirect source of freshwater.[12] This links freshwater regulation with broader climate policies and global warming. In other words, water is both a local as well as a global issue and both dimensions are intrinsically linked.[13]

Overall, water is only available in limited quantities. The gross amount available for human and ecosystem use can be further influenced by other factors, such as pollution that makes water unfit for use, climate change that impacts the local and global water cycle or population growth that impacts per capita availability. Further, global or national-level statistics mask the fact that water availability is context specific since most people rely on water that is accessible in their locality.

The existence of water stress or water scarcity is a function of different parameters.[14] It is a function of overall water use that has, for instance, doubled on a global level between 1960 and 2000.[15] It can also be linked to seasonal variations in water availability, to anthropogenic activity or can relate to forms of control over access which may privilege certain users.[16] In fact, the real problem is not, in most cases, physical availability of water. As acknowledged by the UNDP,

[9] Planning Commission—Government of India, *Eleventh Five Year Plan 2007–12—Volume 1—Agriculture, Rural Development, Industry, Services and Physical Infrastructure* (New Delhi: Oxford University Press, 2008) 44.

[10] United Nations (n 1 above) 133. Cf GC Varughese, 'Water and Environmental Sustainability' in J Briscoe & RPS Malik, *Handbook of Water Resources in India—Development, Management and Strategies* (New Delhi: The World Bank and Oxford University Press, 2007) 184, estimating it at 30%.

[11] Ground Water Management and Ownership—Report of the Expert Group (New Delhi: Government of India, Planning Commission, 2007) 4.

[12] United Nations (n 2 above) 10.

[13] For the World Bank perspective on this point, ch 3.C.2, pp. 92–3.

[14] eg BR Johnston, 'The Commodification of Water and the Human Dimensions of Manufactured Scarcity' in L Whiteford & S Whiteford (eds), *Globalization, Water, and Health—Resource Management in Times of Scarcity* (Santa Fe: School of American Research Press, 2005) 133.

[15] Vörösmarty (n 4 above) 175.

[16] L Mehta, Water for Twenty-First Century—Challenges and Misconceptions (Sussex: Institute of Development Studies, Working Paper 111, 2000) 4.

'the scarcity at the heart of the global water crisis is rooted in power, poverty and inequality'.[17] Water scarcity is nevertheless of primary importance in current policy debates since it constitutes the premise for most of the reforms that are proposed in the water sector. It is also a problematic notion. Firstly, the World Bank bases its call for better managing water in India on existing 'serious water constraints'. Yet, at the same time it readily acknowledges that India is not really water scarce and that it compares quite favourably with a number of other countries.[18] Secondly, population growth is often put forward as one of the major causes of the increasing water scarcity the world and specific countries are facing. Yet, in reality, the growth in population affects disproportionately the poorer segments of society. These are the same people who either do not have access to water in sufficient quantities or use disproportionately much less water per person. Thirdly, there is no agreement among different actors on the kind of water crisis we face and the responses that need to be given.[19]

The notion of water scarcity must also be examined in the context of the different types of water uses. Indeed, in a context where water scarcity is used as premise for a specific set of reform measures, the low water requirement for domestic use is an important consideration. In India, domestic water use in a high population growth scenario is estimated at no more than 6 per cent of available water by 2050.[20] Even if these figures are revised in the future, they make it abundantly clear that India will not face for many decades any physical water scarcity that would make it difficult or impossible to ensure the realization of the human right to water. In that sense, water stress needs to be evaluated at a much more individualized level. Indeed, any individual experiences water stress if s/he does not get access to, say 150 litres per capita per day (lpcd). Millions of people thus experience water stress on a daily level but this is very different from India, or any other country, suffering from water stress in aggregate. The two thus need to be carefully distinguished.[21]

The question of water availability also needs to be examined in a broader context. Firstly, aggregate figures do not specifically indicate what or who makes the water available. While at a global level availability can be ascribed to the overall water cycle, at a more local level it is ecosystems that ensure availability for nature and human needs. It has, for instance, been estimated that forest ecosystems serve as the source areas of 57 per cent of the total freshwater runoff.[22]

[17] United Nations Development Programme, *Human Development Report 2006—Beyond Scarcity: Power, Poverty and the Global Water Crisis* (New York: UNDP, 2006) 2.

[18] World Bank, India—Water Resources Management Sector Review—Inter-sectoral Water Allocation, Planning and Management (Report No. 18322, 1998) 2.

[19] R Iyer, *Towards Water Wisdom: Limits, Justice, Harmony* (New Delhi: Sage Publications, 2007) 42.

[20] National Commission for Integrated Water Resource Development Plan, Report (New Delhi: Ministry of Water Resources, 1999) 155.

[21] cf RR Iyer, 'Water: A Critique of Three Concepts' (2008) 48/1 *Economic & Political Weekly* 15, 16.

[22] Vörösmarty (n 4 above) 167.

Secondly, human activity also contributes to water availability. Thus, large dams boost water availability for human needs by intercepting about 40 per cent of the water that flows off the continents and into oceans or inland seas.[23] In the current scenario where a number of big rivers are already dammed, attention has turned towards inter-basin transfers. The rationale is the same as for traditional dams but the consequences are more significant because the source basin is deprived of part of its natural flow.[24] Both traditional dams and inter-basin transfers are not unequivocally positive in terms of water availability. Indeed, the additional supply for human needs is compensated by the fact that downstream ecosystems suffer from reduced water flows and overall water availability. Additionally, a number of dams cause unjustifiable socio-economic consequences, such as loss of land and livelihood that must be taken into account in assessing them.

Thirdly, total water availability can be influenced by human intervention. Desalinization is one method that contributes to artificially increasing the amount of freshwater by turning unusable water into freshwater. This is technically feasible and has been widely implemented, especially in some very arid countries. Yet, there are potentially negative environmental side-effects linked to the brine waste and it remains much more expensive than traditional sources of water.[25]

A. Socio-Economic Context

The availability of water in general and in specific contexts is a primary issue in the development of laws and policies related to water. Yet, water is so important to human life that availability is only one of a variety of important elements that condition water regulation. Water availability does not necessarily translate into access to water for a specific person in a specific location. Similarly, the fact that the same water source is the basis for different uses gives sectoral allocation an important role in the development of regulatory frameworks. More broadly, poverty is of primary importance in conceiving equitable water regulatory frameworks since the poor are disproportionately affected by insufficient access to water for domestic and livelihood uses.

1. Access to water

The availability of sufficient water at an aggregate or local level is not necessarily an immediate indication of the number of people who have access to sufficient water of adequate quality. Indeed, while there is enough water overall, it was

[23] Vörösmarty (n 4 above) 183.
[24] National Commission for Integrated Water Resource Development Plan, Report (New Delhi: Ministry of Water Resources, 1999) 199 discusses some of these issues.
[25] Vörösmarty (n 4 above) 172.

estimated that in 2004 about 1.07 billion people did not have access to sufficient clean water.[26] Many more do not have access to sanitation. While 79 per cent of the population of all developing countries is deemed to have access to an improved water source, only 49 per cent have access to sanitation.[27] The gulf between access to water and sanitation is even bigger in the case of India where 86 per cent of the population is deemed to have access to water while only 33 per cent has access to sanitation.[28] There are close links between the two, which include the potential impacts on water sources used for domestic uses and the health consequences of the contamination of clean water. The most worrying aspect of these figures is that an overwhelming majority of the people who do not have access to clean water and sanitation are the poorest.[29] Lack of access to water and sanitation is compounded by the fact that it is the direct or indirect cause of at least 1.7 million deaths a year worldwide, again disproportionately affecting the poor.[30]

Water is not only unevenly distributed in time and space but also unevenly distributed among socio-economic groups.[31] In fact, deep inequalities in access to water are visible in a number of contexts. In India, policies suggest that urban dwellers should be provided more water than rural inhabitants, thus causing concern from a non-discrimination point of view.[32] In rural areas, inequalities in access to water often run alongside inequalities in access to land. This is partly linked to the fact that rights of access to water are in many cases partly linked to control over land.[33] Further, there are also social inequalities. In India, caste and religion can, for instance, be important factors in influencing access to water. Thus, in a study in the district of Allahabad in Uttar Pradesh, it was found that while a total of 50 per cent of open castes had access to safe sources of water, only 5 per cent of scheduled tribes and 16 per cent of other backward castes had access to a safe source.[34]

Basic figures concerning access mask a more dire reality. Indeed, all the statistics tell us is that nearly 1.07 billion people worldwide do not have 'reasonable access' to water which means that they do not even have access to 20 litres per person per day at a distance of not more than 1 kilometre from the dwelling.[35] This is a deeply conservative estimate of water needs which is not an appropriate measure of access to water. Statistics also tell us that another

[26] UNDP (n 17 above) 300, 308.
[27] UNDP (n 17 above) 308.
[28] UNDP (n 17 above) 307. See ch 5.A for a more elaborate discussion of these figures.
[29] UNDP (n 17 above) 7.
[30] United Nations (n 2 above) 508.
[31] United Nations (n 1 above) 46.
[32] ch 5.A.2–3 for the drinking water policy context and ch 6.D.1, pp. 204–6 concerning discrimination.
[33] ch 2.B.3–4.
[34] FF Rizvi & S Thorat, Dalit and Water—Reflection on Accessibility and Discrimination (Paper presented at the IWMI-Tata Water Policy Programme Annual Partners' Meet, Gujarat, 2006).
[35] World Health Organization and United Nations Children's Fund, Global Water Supply and Sanitation Assessment 2000 Report (2000).

1.6 billion people only have 'reasonable access' or, as defined in some other documents, a 'basic level of service'.[36] This level of access does not include access through unprotected wells, unprotected springs, rivers or ponds, vendor-provided water, bottled water or tanker truck water but includes a variety of other forms of access such as public standpipe, borehole, protected dug well, protected spring, rainwater collection and household connection.[37] On the whole, this means that 2.67 billion people in the South (or about 52 per cent of the population) have access at best to 20 litres per person per day within a kilometre of their house.[38]

2. Multiple uses of water

Water has a multiplicity of uses. Its overall consumption has tremendously increased over the past two centuries with a fifteenfold increase in global water withdrawals between 1800 and 1980.[39] All uses rely on the same overall limited supply available at any one point. As a result, these need to be individualized and ranked in law and policy instruments according to their importance. This is particularly true where different uses of water may conflict with each other.

Uses of water include a number of different activities from abstraction of water from a source and diversion of the flow of a river to storage, as well as activities such as disposing of waste in a manner which may detrimentally impact on a water resource. All these uses can take place in a number of different contexts, some of the main ones being social uses, agriculture and economic development. The latest figures for different water uses in India are 86.5 per cent for irrigation, 8.1 per cent for domestic water and 5.4 per cent for industry for the year 2000.[40] Estimates of the Planning Commission for 2010 confirm these figures but indicate that the shares of industrial and domestic uses are likely to increase in the future.[41] In fact, the World Bank's estimates in the late 1990s were for a doubling of domestic water use and a whopping 340 per cent increase for industrial use by 2025.[42] These figures must be understood as estimates.

[36] World Health Organization, The Right to Water, Doc. WA 675 (2003) 12.

[37] UN Department of Economic and Social Affairs, Progress Towards the Millennium Development Goals, 1990–2005—Goal 7—Ensure Environmental Sustainability (2005) 21, available at <http://unstats.un.org/unsd/mi/goals_2005/Goal_7_2005.pdf>.

[38] Data concerning India is introduced in ch 5.A.

[39] Vörösmarty (n 4 above) 174.

[40] FAO, AQUASTAT database, available at <http://www.fao.org/nr/water/aquastat/main/index.stm> (latest figures available are for 2000).

[41] Planning Commission (n 9 above) 46 and UA Amarasinghe et al., India's Water Future to 2025–2050: Business-as-Usual Scenario and Deviations (Colombo: International Water Management Institute, 2007) 9.

[42] World Bank, India—Water Resources Management Sector Review—Inter-sectoral Water Allocation, Planning and Management (Report No. 18322, 1998) 9.

Indeed, the eleventh plan estimates, for instance, that drinking water demand is only 1 per cent of overall demand.[43]

The first category of uses encompasses domestic and livelihood water uses. The primordial use is drinking water which includes not only a minimum required quantity but also specific quality, which makes it harmless in health terms. Other domestic uses such as cooking, bathing, ablutions and house cleaning also constitute basic uses. These uses are increasingly met from groundwater, especially in rural areas where as much as 80 per cent of domestic water supply comes from groundwater.[44] The most recent widespread survey indicates that for 55.1 per cent of villages the major source of drinking water is handpumps or tubewells followed by taps for 18.5 per cent of the rural population and wells for 17.6 per cent. These overall figures mask significant inter-state differences as only 4 per cent of the rural population accesses water through taps in Uttar Pradesh while it is 45 per cent in Maharashtra.[45]

Beyond strictly domestic uses, water has a number of livelihood uses, such as water necessary to grow one's own food, a central issue for the majority of people living in rural areas in the South. It also comprises water used in activities, which are basic livelihood activities, such as the abstraction of water for subsistence crops or basic food crops used by people who do not grow their own food. Further, there is also water used as part of a livelihood as in the case of fisher and washer people, riverbed farmers or potters. In addition, there are activities not directly based on water such as cattle rearing that require water as a part of the activity. This latter use is, for instance, specifically recognized as a special water use in India where an entitlement of 30 litres per day per capita is granted to animals in specific areas such as hot or cold deserts.[46] As with other water uses, livelihood uses are increasingly groundwater based.[47]

The second category refers to ecological uses of water. This is of increasing importance in the context of increased demand for water and diminishing per capita availability. At the most basic level, water brings life to all ecosystems. Diversions of rivers or abstraction from groundwater aquifers for irrigation, industrial use or domestic needs not only have social implications but also direct ecological consequences. Ecological uses of water refer in general to the quantity of water that is necessary to sustain existing ecosystems. They are important in themselves and also have impacts on other uses. Indeed, in dramatic situations like the Aral Sea that has been deprived of most of its feeding rivers, the 'costs' of the diversion of the river flow for agricultural activities upstream have impacts not only on the Aral

[43] Planning Commission—Government of India, *Eleventh Five Year Plan 2007–12—Volume II—Social Sector* (New Delhi: Oxford University Press, 2008) 162.

[44] United Nations (n 2 above) 78.

[45] National Sample Survey Organisation, Report on Village Facilities, NSS 58th Round (July–December 2002) (New Delhi: Ministry of Statistics and Programme Implementation, 2003) 24.

[46] Accelerated Rural Water Supply Programme (ARWSP) Guidelines, 1999, s 2(2)(2).

[47] NK Dubash, 'The Electricity-Groundwater Conundrum: Case for a Political Solution to a Political Problem' (2007) 47/52 *Economic & Political Weekly* 45, 46.

Sea but also on human uses of water for people inhabiting neighbouring areas, such as fisher people having lost their livelihood.[48]

The third category, agriculture, is the most important in terms of quantity of water used throughout the South, accounting for more than 80 per cent of all water uses.[49] Most of this water goes to irrigated agriculture, which has contributed an increasingly important percentage of agricultural output over the past few decades.[50] A growing share of agricultural water is pumped from the ground thanks to the increasing availability of pumping devices and in India, as much as 70 per cent of national agricultural production is now supported by groundwater.[51] The increase in the use of groundwater for irrigation has been rapid. Thus, tubewell irrigation increased from less than 1 per cent of the irrigated area in 1960 to a third by the end of the 1990s.[52] In fact, most of the increase in the net irrigated area in recent years has come from groundwater.[53] Overall, more than 50 per cent of irrigation in India relies today on groundwater.[54] While agriculture is the biggest overall user of water as a sector, this general statement masks huge variations. On the one hand, irrigated agriculture represents on the whole a small percentage of land under cultivation.[55] On the other hand, some of the most water-intensive crops are, like cotton, cash crops whereas crops like millets that feed an important percentage of the population are often grown under rain-fed conditions.

The fourth category can be loosely defined as use of water for economic development. This includes water use by industries, as well as water for transport, hydropower and other large-scale activities fostering economic growth. In India, industry uses account for a relatively low percentage at present compared to the 22 per cent at a global level.[56]

In addition to the above-mentioned water uses, virtual water is also increasingly important. The notion of virtual water highlights the fact that even where water is not actually used in a specific place, indirect water use can take place.[57] Virtual

[48] eg World Bank, Saving a Corner of the Aral Sea, available at <http://go.worldbank.org/UI222F8DY0>.

[49] UNDP (n 17 above) 138. Note that the balance is different in OECD countries where agriculture and industry each use more than 40% of available water (ibid).

[50] A Gulati, R Meinzen-Dick & KV Raju, *Institutional Reforms in Indian Irrigation* (New Delhi: Sage, 2005).

[51] United Nations (n 2 above) 78.

[52] A Narayanamoorthy & RS Deshpande, *Where Water Seeps!—Towards a New Phase in India's Irrigation Reforms* (New Delhi: Academic Foundation, 2005) 62.

[53] J Briscoe & RPS Malik, *India's Water Economy—Bracing for a Turbulent Future* (New Delhi: The World Bank and Oxford University Press, 2006) 8.

[54] National Commission for Integrated Water Resource Development Plan, Report (New Delhi: Ministry of Water Resources, 1999) ii and Dubash (n 47 above) 46.

[55] In fact, 68% of the total net sown area is rainfed agriculture. eg Showcasing Rainfed Agricultural Technologies of Southern Tamil Nadu (Madurai: DHAN Foundation, 2007).

[56] United Nations (n 2 above) 509.

[57] On virtual water, eg T Allan, 'Watersheds and Problemsheds: Explaining the Absence of Armed Conflict Over Water in the Middle East' (1998) 2/1 *Middle East Review Intl Affairs* 49.

water constitutes a useful way to take into account the water implications of different activities. It provides a tool to assess indirect water transfers between rural and urban areas. From this perspective, while it is rural areas that use most water because irrigation is the main user of water, the overall water use balance is less skewed because cities consume an increasingly important part of the crops grown under irrigation.

This survey of some of the main types of water uses does not reflect the tensions and contradictions that may surface between different water uses. Given the limited availability of water in specific contexts and at specific times, the different uses of water have the potential to be in competition with each other. Different types of competing uses can arise. In a rural setting, the two main uses of water are domestic use and irrigation. In situations where water comes primarily from underground sources, availability is limited by the depth of wells or other pumping mechanisms, by the movements of the water table and by the cost of energy required to lift water in situations where it cannot be accessed through manually operated equipment. Since all water uses depend on the same sources of water, excessive pumping for irrigation has the potential to reduce drinking water availability while less irrigation also has impacts on the success of the crops and therefore on food sovereignty. Indeed, in the case of India, 85 per cent of drinking water needs are now met from groundwater but this accounts only for 5 per cent of all groundwater extraction.[58] Similarly the price of electricity for pumping can have devastating impacts on people's lives if it makes it unaffordable to pump enough water for domestic and subsistence uses.

Another issue arises where available water is contaminated. Industries that pollute rivers or groundwater do not directly reduce water availability but restrict the water that can be used as drinking water. Another type of conflict between different uses is illustrated by the case of dams that restrict downstream flows to the extent that this allows salt ingress in the parts of the river close to the estuary. In a dramatic case from Pakistan, salt ingress due to reduced river flows has been identified up to 54 kilometres from the mouth of the river.[59] This directly affects the environment as well as livelihood since salt ingress threatens existing crops grown in the area.

Competition between different uses can also be identified by examining water uses from an urban-rural perspective. Transfers from rural to urban areas will be an increasingly significant policy challenge in the context of rapid urbanization. The tensions that arise between urban and rural areas derive from the fact that it is the same limited supply of water that is used for all purposes. Where water used for agriculture is diverted for the supply of domestic water to cities, important policy questions arise because this transfer cannot be simply analysed as a transfer from

[58] eg KV Raju, Sustainable Water Use in India: A Way Forward (Paper Commissioned by the NCAP–ISEC and supported by ADB as part of the Policy Research Networking to Strengthen Policy Reforms in Agriculture, Food Security, and Rural Development, 2004).

[59] S Husain, 'Fated to Salt' (2007) 15/16 *Down to Earth* 44.

agriculture to domestic use. There is first a marked difference between water used to grow subsistence food crops or cash crops. Further, the transfer has impacts on the environment which may be more or less severe or permanent in cases where water withdrawals go beyond the recharge potential of the area. This impacts not only the flow of rivers but also their potential for recharging groundwater in downstream areas, their cleansing capacity and the availability of water to downstream users. Other issues that arise include situations where dams are constructed in rural areas, entailing impacts for local populations from displacement to lost livelihood, with the aim of producing electricity mostly used in cities and to send drinking water to other areas. These include both indirect and direct transfers of water from relatively poorer rural areas to relatively wealthier urban areas.

Finally, the question of water use also needs to be approached from the point of view of the use of different bodies of water. While it is increasingly acknowledged that water needs to be addressed as a unitary resource, from a practical as well as from a law and policy perspective a distinction between different water uses is required because current laws and policies are often sectoral. One of the main questions that arise concerns the distinction between surface and groundwater. Indeed, uses of surface and groundwater in many countries have rapidly changed in the past few decades. The development of new technologies to pump groundwater has had immense impacts on consumption patterns. In India, for instance, while the net irrigated area from canals increased 2.1 times from 1950–51 to 1999–2000, the area irrigated by groundwater increased 5.6 times.[60] In other words, the importance of groundwater as a source of water for irrigation increased massively during this period from 29 to 59 per cent. Groundwater has thus been identified as the main factor behind agricultural growth in Punjab and Haryana, two of the leading states in terms of agricultural production.[61]

3. Poverty and water

Poverty has been defined as a 'sustained or chronic deprivation of resources, capabilities, choices, security and power'.[62] Absolute poverty is in turn characterized by severe deprivation of essential human needs. The first four needs mentioned by the Programme of Action of the World Summit for Social Development are food, safe drinking water, sanitation facilities and health.[63] This

[60] Narayanamoorthy & Deshpande (n 52 above) 260.

[61] S Dharmadhikary, *Unravelling Bhakra—Assessing the Temple of Resurgent India* (Badwani: Manthan, 2005) 113.

[62] Draft Guiding Principles 'Extreme Poverty and Human Rights: The Rights of the Poor', Resolution 2006/9, Implementation of Existing Human Rights Norms and Standards in the Context of the Fight Against Extreme Poverty, Report of the Sub-Commission on the Promotion and Protection of Human Rights on its Fifty-Eighth Session, UN Doc. A/HRC/2/2-A/HRC/Sub.1/58/36 (2006), principle 1.

[63] Programme of Action of the World Summit for Social Development, in Report of the World Summit for Social Development, UN Doc. A/CONF.166/9 (1995) 41.

confirms the primary importance of water in debates on poverty since it is directly involved in two of the above-mentioned needs and is a key component of the realization of the other two. Poverty eradication has been a primary policy goal of the international community for a number of years. Thus, in 1995, states committed themselves to 'eradicating poverty in the world, through decisive national actions and international cooperation, as an ethical, social, political and economic imperative of humankind'.[64]

The central role of water in poverty eradication gives the latter a key role in the development of water law for the twenty-first century. Indeed, access to sufficient safe water for drinking, domestic and livelihood uses is a pre-condition for a life of dignity and free of water-borne diseases, a primary killer that disproportionately affects the poor. Impacts of poor quality water on health and life are sobering. Inadequate water, as well as the directly related issues of inadequate sanitation and hygiene, is the cause worldwide of a staggering 1.8 million deaths due to diarrhoeal diseases and 1.3 million deaths due to malaria, the two main killers in this category.[65] This mortality affects nearly exclusively children with 90 per cent of these deaths being of children under 5 years of age. Additionally, according to the Disability-Adjusted Life Year (DALY) measure, 62 and 46 million healthy life years are lost annually to diarrhoeal diseases and malaria each year respectively.[66] The problems that arise can be classified in different categories. Some diseases are directly linked to the lack of access to safe drinking water, itself related to ecosystem conditions as well as human interventions such as the direct dumping of sewage into surface waters. Other diseases are associated with ecological conditions that favour disease vector breeding. These may be natural or anthropogenic such as where irrigation systems, dams or urban water systems are improperly planned.[67] The only constant factor is that it is always the poor who bear an overwhelming share of this morbidity.

Water is also one of the primary causes of poverty and impoverishment. Thus, in certain cases access to water for one person can lead to someone else's impoverishment. This is, for instance, the case of a dam built to provide drinking water to unserved urban or rural dwellers that displaces, and simultaneously impoverishes, the people that are uprooted by the dam. A real life example is that of the Sardar Sarovar Dam.[68] The twin dimensions of water as a vector in poverty eradication and impoverishment highlight its key role. This has been confirmed at the international level by the UNDP which sees the existing water crisis as being above all a crisis for the poor.[69]

[64] Copenhagen Declaration on Social Development, in Report of the World Summit for Social Development, UN Doc. A/CONF.166/9 (1995) 13.
[65] United Nations (n 1 above) 209.
[66] Ibid.
[67] Vörösmarty (n 4 above) 195.
[68] eg Tata Institute of Social Sciences, Performance and Development Effectiveness of the Sardar Sarovar Project (Mumbai: TISS, 2008).
[69] UNDP (n 17 above) 7.

The importance of putting poverty eradication at the centre of water law is highlighted by the fact that the poor have been regularly blamed either for their own situation or for causing unsustainability.[70] In fact, it is the poor that disproportionately suffer from the negative impacts of access to insufficient safe water. Further, the poor often suffer the consequences of the lifestyles of richer people who use more water and other environmental resources but often manage to avoid the negative consequences of these activities.[71] The poor thus tend to suffer the negative consequences of benefits they have not enjoyed in the first place. The facts speak for themselves but this is not yet systematically recognized in law and policy making.

B. Law and Policy Context

Water law in India has been and is an evolving field of law. Rules of access to and control over water for drinking and irrigation have played a defining role in the evolution of human civilization. Yet formal modern water legislation, whose start dates back to the nineteenth century, has not been given the central importance that it deserves despite the importance of water for human life and life on earth. Some of the possible explanations include the fact that water itself was seen as being beyond human appropriation. Another factor is that the availability of freshwater in a given locality was seen as largely non-negotiable since the technological options to transport water, make freshwater out of saline water or dig deeper to mine aquifers were comparatively few. The tremendous changes that have occurred during the twentieth century with the introduction of big dams on a large scale, new technologies to mine groundwater from ever deeper layers with a concomitant increase in the use of groundwater or desalinization techniques have fundamentally altered the basic premises on which formal water law, and the understanding of water, have been built. This has led to a rethinking of water law and policy which first crystallized in the adoption of water policies and has more recently led to the introduction of a spate of water legislation. As chapters 3 and 4 make it clear, the reforms that have and are taking place reflect one specific understanding of the challenges that the water sector is facing. As a result, they do not necessarily address all the relevant aspects of water regulation that require new regulatory measures. Some of the gaps identified in ongoing reforms are addressed in chapter 6.

[70] The eviction of people living close to the Yamuna river bed in Delhi was, for instance, justified by the fact that their very presence was an important cause of the river's pollution. *Wazirpur Bartan Nirmata Sangh v. Union of India* CM Nos. 11672-73/2006 in WP(C) No. 2112/2002 (High Court of Delhi, 2006).

[71] United Nations Environment Programme, *Global Environment Outlook—GEO4—Environment for Development* (Nairobi: UNEP, 2007) 315.

1. Context for water regulation

From a legal perspective, water can be looked at from two distinct perspectives. On the one hand, it can be seen as one of many substances whose regulation has and is evolving in line with physical and social scarcity. On the other hand, it can be approached as something that has always fallen into a separate category because of its direct contribution to sustaining life on earth. This dichotomy—and tension—partly explains why water law remains a separate area of the law that has its specific rules and principles. Yet, while it is easy to understand why water law would have evolved as a separate field in the past, it is imperative to recognize today that there are an increasing number of links with other areas, from human rights to environmental concerns.

The regulation of water is a difficult task because of the nature of water and the multiplicity of uses to which it is put. Water law must take into account these different dimensions, prioritize uses and address issues concerning the links between water and other fields such as food, health, agriculture, and energy.

Firstly, water regulation needs to take into account the links between drinking water needs, food-related water use, which include the relatively minimal amounts needed for cooking food, other domestic water needs as well as the links between water availability and quality, and diseases.

Secondly, water regulation needs to take into account not only the direct links between water availability and food security but also links between water uses by humans, water needs of domesticated species and water needs of other species as well as ecosystems in general. This balance between social, environmental, and economic needs is complex and controversial. As a result, while environmental law addresses certain water-related needs of species, such as in the case of the Convention on Wetlands of International Importance Especially as Waterfowl Habitat or water pollution as in the case of the Water Act 1974,[72] water law has only recently started to address the broader connections between water and the environment.

Thirdly, water regulation must address the different though complementary water needs of cities and rural areas. This includes different issues. Urban areas often draw a lot of water from outside of their immediate ecosystem because the population density is too high for the area concerned. As a result, water is brought in from neighbouring or distant areas, creating imbalances and possible scarcity in areas which may in themselves be abundantly provided with water sources. Similarly, while rural areas consume most of the irrigation water, from a virtual water perspective, this is water which is also largely consumed by cities while millions of rural people may be struggling to meet their basic domestic water needs.

[72] Convention on Wetlands of International Importance Especially as Waterfowl Habitat, Ramsar, 2 February 1971, 996 UNTS 245 and Water (Prevention and Control of Pollution) Act 1974.

Fourthly, laws need to take into account issues from the local level up to transboundary aspects of water use. The importance of water for basic domestic needs, livelihood, subsistence activities, industrial, and other economic activities is similarly acute in most states and in most countries. As a result, given that most river basins span several countries, the allocation of available water at the international level is also of primary importance in the context of national water law.

2. International water law context

International water law is characterized by a basic dichotomy which influences its relevance in the context of this book. On the one hand, it is a relatively old area of international law that includes a series of well-developed principles. These principles have, at least to a certain extent, permeated the legal systems of most countries over time. As a result, some of the basic principles of international water law are relevant in the analysis of domestic water law. On the other hand, international water law remains an underdeveloped area of the law with a relatively limited scope. Water law at the national level is more comprehensive in scope and content. Consequently, international water law is only of limited relevance in trying to conceive new water laws able to address the challenges that countries face in the twenty-first century.

Existing international water law largely focuses on transboundary waters. It has evolved over time from being primarily concerned with rules concerning navigation to progressively include non-navigational uses of transboundary watercourses.[73] One of the major contributions of international water law has been to gradually allow states to define principles concerning the sharing of water that neither unfavourably benefit the upstream nor the downstream state. Yet, the fact that some states are predominantly upstream or predominantly downstream states is one of the factors that has hampered the development of a binding regime at the international level and its coming into force.

Indeed, one of the main difficulties has been that each state seeks to exert as much control over waters found in their territory as possible. Qualifications have been introduced over time in recognition of a necessity to cooperate and collaborate on certain water-related issues. One of the first aspects where cooperation was accepted among countries sharing a river basin concerned the necessity to allow all riparian states to use a river for navigation purposes throughout the course of the river. This was confirmed in a decision of the Permanent Court of International Justice where the court found that there was a community of interests among all riparian states to facilitate navigation along the entire course of the river.[74] The recognition of a community of interests considers the management of

[73] SC McCaffrey, *The Law of International Watercourses* (Oxford: Oxford University Press, 2007) 64.

[74] *Case Relating to the Territorial Jurisdiction of the International Commission of the River Oder* Series A, No. 23 (Permanent Court of International Justice, 1929).

international watercourses as an integrated whole. This leads to the idea of joint management but this has not received widespread support in practice.[75]

With regard to the use and diversion of the water by one state, it took much longer for all states to agree to sharing principles, since upstream states were not inclined to part with control over what was considered their water. Even though there is still no overall consensus on the modes of cooperation, there has been a significant evolution in the position advocated by states. Two main theories were proposed in earlier times by states seeking to assert full control over the waters flowing through their territories. The Harmon doctrine, proposed that states enjoy absolutely sovereignty over the water found within their territories and should be free to dispose of the water in ways they see fit.[76] As opposed to the Harmon doctrine favoured by upstream states, some downstream states proposed that each lower riparian state should have the right to a full flow of water of natural quality.[77]

Assertions of sovereign rights have progressively been watered down due to the realization of most states that it did not encourage cooperation with other states.[78] Indeed, the states that attempted to claim absolute rights failed to prevail.[79] Nevertheless, this evolution has taken time and is still ongoing. A case in point is the Lake Lanoux case, which did not condone the French claims of full sovereignty but stopped at requiring notification to the downstream state and the safeguarding of Spain's own rights in the watercourse.[80]

The main international treaty concerning non-navigational uses of watercourses is the UN Water Convention.[81] The basic principle the UN Water Convention proposes for using international watercourse water is equitable and reasonable use.[82] It provides a balance between the competing interests of different groups of states.[83] The basis for watercourse use is therefore agreement

[75] M Fitzmaurice & G Loibl, 'Current State of Development in the Law of International Watercourses' in SP Subedi (ed), *International Watercourses Law for the 21st Century—The Case of the River Ganges Basin* (Aldershot: Ashgate, 2005) 19.

[76] J Bassett Moore, 'Case of the Rio Grande', *A Digest of International Law* (Washington: Government Printing Office, Vol. 1, 1906) 653.

[77] eg M Fitzmaurice, 'General Principles Governing the Cooperation Between States in Relation to Non-Navigational Uses of International Watercourses' (2005) 14 *Ybk Intl Environmental L 2003* 3, 7.

[78] PW Birnie & AE Boyle, *International Law and the Environment* (Oxford: Oxford University Press, 2002) 301.

[79] L Caflish, 'Règles générales du droit des cours d'eau internationaux' (1989) 219 *Recueil des cours —Académie de droit international* 2.

[80] *Lac Lanoux Arbitration (France v Spain)* 24 *Intl L Rep* 101 (1957).

[81] Convention on the Law of the Non-navigational Uses of International Watercourses, New York, 21 May 1997, UN Doc. A/51/869. The main other treaty is the Convention on the Protection and Use of Transboundary Watercourses and International Lakes, Helsinki, 17 March 1992, UN Doc. ENWA/R.53. It is now open for universal membership but is not covered here since it was negotiated as a regional agreement and has not been ratified by any non-UNECE state to-date.

[82] Convention on the Law of the Non-navigational Uses of International Watercourses, New York, 21 May 1997, UN Doc. A/51/869, art 5.

[83] eg J Sohnle, *Le droit international des ressources en eau douce: Solidarité contre souveraineté* (Paris: La Documentation française, 2002).

among concerned states concerning their respective needs. The factors that make up equitable and reasonable use are partly listed in the convention. These include natural factors such as geographical, hydrological, climatic, and ecological aspects; socio-economic factors such as the social and economic needs of the watercourse states concerned and of the population dependent on the watercourse in each watercourse state; the effects of the use or uses of the watercourses in one watercourse state on other watercourse states; existing and potential uses of the watercourse; conservation, protection, development and economy of use of the water resources of the watercourse and the costs of measures taken to that effect; and the availability of alternatives, of comparable value, to a particular planned or existing use.[84]

The breadth of factors included under the convention implies that they may in some circumstances conflict with each other. However, it does not prioritize among factors and simply indicates that a specific weight has to be given to each factor in view of its importance in a specific case.[85] In other words, while the principle of equitable and reasonable use now plays a central role in international water law, its specific content remains a matter of debate and a question which will have to be addressed again in the future by states. It has thus been criticized for not being capable of a precise definition.[86]

Under the convention, equitable and reasonable use is balanced with the principle of non-significant harm. In fact, the final text is a package deal that balances equitable utilization favoured by upper riparian states and the no harm principle favoured by downstream riparian states.[87] One of the implications of this compromise is that environmental interests have not become predominant in the existing international water law regime and can be trumped by considerations of equitable utilization. Sustainability can thus be compromised.[88] Indeed, while there was substantial debate concerning the place of environmental aspects and sustainability, the principle of sustainable use has not been adopted as a principle that would override equitable and reasonable utilization.[89] Additionally, the convention does not stop member states from adopting agreements which may depart from the provisions of the Convention, hence potentially further restricting the impact of the compromise reached from an environmental and sustainability perspective.[90]

[84] Convention on the Law of the Non-navigational Uses of International Watercourses, New York, 21 May 1997, art 6.
[85] ibid art 6(3).
[86] Birnie & Boyle (n 78 above) 303.
[87] Fitzmaurice & Loibl (n 75 above) 26.
[88] A Hildering, *International Law, Sustainable Development and Water Management* (Delft: Eburon, 2004) 57.
[89] eg Birnie & Boyle (n 78 above) 307 and P Wouters, *The Legal Response to International Water Scarcity and Water Conflicts—The UN Watercourses Convention and Beyond* (University of Dundee: Water Law and Policy Programme, 2003) 20.
[90] Fitzmaurice & Loibl (n 75 above) 29.

The adoption of the convention was in itself a landmark development since it took negotiators many years to approve this text. Nevertheless, the difficulties encountered in negotiating it are reflected in the fact that its scope is relatively limited. Thus, it only applies to international watercourses and is therefore not a convention addressing freshwater in general. Further, its operative principles are relatively outdated as it fails to break clearly with the traditional principle of equitable and reasonable use in favour of a sustainability based approach.[91] Additionally, since the convention does not affect pre-existing agreements, it does not affect agreements that may not be compatible with its principles.[92] While the convention does not break much new ground at the conceptual level, states, including India, have proved unwilling to ratify it. This confirms that freshwater remains an issue over which states are fearful of losing control. As a result, even relatively weak coordination measures appear threatening to many.

a) International and national water law

In general, international water law is well developed with regard to cooperation among states concerning issues and activities that are clearly transboundary in scope such as navigation on international watercourses. In recent decades, the importance of collaboration on non-navigational aspects of international watercourses has rapidly grown and is now recognized as a core objective of international water law. Recent instruments also seek to move beyond the artificial divide between surface and groundwater. Further, international water law has, for instance, been influenced by the notion of sustainable development.[93] There is also a recognition of the need to integrate an environmental perspective even if much more needs to be done.

Yet, international water law remains of limited relevance when addressing water regulation at the national level. Firstly, existing conventions focus largely on international watercourses and transboundary impacts. International water law thus fails to provide an effective basis for allocating and protecting water and is not developed with regard to cooperation on issues related to water found within national boundaries.[94] Secondly, even the limited UN Water Convention has failed to come into force more than a decade after its adoption. Thirdly, international water law has failed to effectively move beyond issues of control over water. Thus, the social and human right dimensions of water remain largely absent. While the adoption of General Comment 15 of the International Covenant on Economic, Social and Cultural Rights or other treaty recognition of the

[91] Hildering (n 88 above) 51.

[92] SP Subedi, 'Regulation of Shared Water Resources in International Law: The Challenge of Balancing Competing Demands' in SP Subedi (ed), *International Watercourses Law for the 21st Century—The Case of the River Ganges Basin* (Aldershot: Ashgate, 2005) 7.

[93] Sohnle (n 83 above) 255.

[94] Hildering (n 88 above) 58.

human right to water provide a water perspective to human rights law,[95] this only constitutes a conceptual framework, which needs to be integrated in water law instruments if it is to be effectively implemented in the practice of member states.[96] Under the UN Water Convention, even though interpretation of the different provisions would most probably lead to giving priority to vital human needs, there is no inherent priority for drinking water.[97]

This poses broader questions concerning the links between international and national water law, in particular in the context of ongoing reforms. This can be best compared to the situation in environmental law. The latter has developed rapidly over the past few decades and has largely developed in tandem at the national and international levels. There is thus readily identifiable cross-fertilization between international and national environmental law. This does not apply to the same extent in the case of water law as national water law shares relatively little in common with international water law compared to environmental law. This is in large part due to the fact that national water law is much more developed than international water law. While the latter focuses to-date mostly on shared watercourses, national water law addresses in most countries of the world a much broader array of issues. Overall, international water law does not contribute much to the development of national water law since it lacks a broad framework that would make it more relevant in the formulation of national laws.

3. International water policy context

The relative underdevelopment of international water law does not imply that international instruments have no relevance or impact on the development of water law. In fact, there exist a multitude of other international, largely non-binding, instruments. There is thus a broad international policy consensus on the shape that domestic water law should take. This policy consensus crafted over the past two decades is noteworthy for its relatively low profile in legal terms but a high degree of compliance. As a result, there is an important international context to ongoing water law reforms in India.

The existing international policy consensus is found in a variety of soft law instruments as well as in the strategies and policies of specific institutions, in particular development banks like the World Bank. One of the key documents in this context is the Dublin Statement on Water and Sustainable Development.[98]

[95] Committee on Economic, Social and Cultural Rights, General Comment 15: The Right to Water (Articles 11 and 12 of the International Covenant on Economic, Social and Cultural Rights), UN Doc. E/C.12/2002/11 (2003).

[96] On the human right to water, ch 2.C.1 and ch 6.C.

[97] Hildering (n 88 above) 100.

[98] Dublin Statement on Water and Sustainable Development, International Conference on Water and the Environment, Dublin, 31 January 1992.

Its four key principles have become the framework of reference to which most proponents of water sector reforms refer. These are:

1) the recognition that water is a finite and vulnerable resource, essential to sustain life, development, and the environment;
2) the need for water development and management to be based on a participatory approach, involving users, planners and policy-makers at all levels based on the principle of subsidiarity;
3) the recognition that women play a central part in the provision, management, and safeguarding of water; and
4) the recognition that water has an economic value in all its competing uses and should be recognized as an economic good. Within this last principle, a basic right of all human beings to have access to clean water and sanitation at an affordable price is recognized.[99]

The last principle is not only the most important in practice but is also the one that most clearly sets out the overall policy agenda of the Dublin Statement. Thus, while principle 1 can be largely understood as proposing a conservationist agenda, the explanatory paragraph of principle 4 makes it clear that the environmental dimension of the Dublin Statement must be understood within the context of the recognition of water as an economic good. Indeed, the need to protect water is ascribed to the failure to put an economic value on water in the past. This indicates that environmental concerns are at least in part a justification for proposing a new conception of water. Additionally, the importance ascribed to turning water into an economic good is highlighted by the fact that it is this change which is expected to foster economic efficiency, socially equitable outcomes as well as water conservation. All hopes are thus pinned on this new economic rationale for the water sector. The Dublin principles have been reiterated in a number of contexts and have, for instance, been substantially incorporated in the policies of the World Bank.[100]

Besides the central role of the Dublin principles, the international community has taken a number of different policy initiatives related to water. One of the central instruments of development policy at present are the Millennium Development Goals (MDGs) adopted in 2000 by the UN General Assembly.[101] These are currently seen as the markers of the development process and poverty eradication. The MDGs are framed as a series of percentage-wise time-bound targets. These include water-related goals. In particular, UN member states have pledged to halve by 2015 'the proportion of people who are unable to reach or to afford safe drinking water'.[102] An assessment of the progress towards the

[99] ibid principles 1–4.
[100] Principles for water sector reforms are analysed in more detail at ch 3.
[101] United Nations General Assembly Resolution 55/2, United Nations Millennium Declaration, 8 September 2000, UN Doc. A/RES/55/2.
[102] ibid para 19.

achievement of these goals by the UNDP showed that, in the case of access to water, if current trends continue up to 2015 only 315 million additional individuals will have access to water. This is a shortfall of 209.9 million individuals towards the fulfilment of the goal that would imply a reduction from 1.036 billion in 2002 to 525 million in 2015.[103] The situation in South Asia is even more dismal since projections imply that only 5 million additional people will be covered by 2015 against a goal of 108.9 million.[104]

The MDGs have, for instance, been reiterated in the Plan of Implementation of the World Summit on Sustainable Development.[105] The Plan provides more specific guidance on their realization, in particular with regard to sanitation. Under the section dealing with natural resources and their importance for economic and social development, it proposes a number of actions to achieve the MDGs. These include the mobilization of financial resources, technology transfer and capacity building, fostering access to information and participation in decision-making related to water, focusing on water pollution prevention and adopting prevention and protection measures to promote sustainable water use and to address water shortages.[106]

Beyond the specific links with the MDGs, the Plan includes a number of water-related policy pronouncements. Under the poverty eradication heading, the Plan recognizes, for instance, the need for concerted action of all countries to prioritize water and sanitation in national sustainable development strategies.[107] It also highlights the links between agriculture and water, recognizes the need to implement combined plans that address water and land and the need for strengthening the role of all actors in monitoring and managing the quantity and quality of land and water resources.[108] The Plan also calls for the introduction of policies and laws that 'guarantee well defined and enforceable land and water use rights and promote legal security of tenure' as well as employing market-based incentives for agricultural enterprises and farmers to monitor and manage water use and quality.[109]

4. National water law framework

Water law has at the same time developed as a separate field of law while sharing a number of links with other areas of law. This gives water law a number of general

[103] UNDP (n 17 above) 44.
[104] ibid.
[105] Plan of Implementation of the World Summit on Sustainable Development, Report of the World Summit on Sustainable Development, Johannesburg, 26 August–4 September 2002, UN Doc. A/CONF.199/20.
[106] ibid para 24.
[107] ibid para 6.
[108] ibid para 38.
[109] ibid para 38(i–j).

characteristics that are directly relevant in the context of ongoing reforms taking place in India.

Firstly, water law has traditionally been made up of different rules for different bodies of water. Surface and groundwater have, for instance, often been treated separately. Further, water has been treated separately in law according to its different uses. Thus, irrigation has been treated separately from drinking water. Similarly water quality issues were for a long time treated mostly as a health issue while water pollution was addressed as an environmental issue. The legacy of this scattered approach to water has been a legal framework that lacks in cohesion and clarity. In India, this is made particularly problematic because of the absence of a framework water legislation either at the state or Union level that would provide overall guidance.

Secondly, water law has often been characterized by a basic conceptual dichotomy. On the one hand, legal systems have recognized that water is unlike other substances. The recognition that water is directly linked to human survival and that it cannot easily be subjected to the usual rules of ownership have been the basis for the special consideration of water in law. Thus, Roman law provided, for instance, that running water was by nature common to all. The public nature of water extended to all rivers, to the use of river banks as well as to the right to fish in rivers.[110] Similar provisions exist until today as in the case of Uruguay where all waters, surface and groundwater, are part of the public domain.[111] On the other hand, different forms of appropriation have been condoned. The state often gets sweeping powers ranging from the position of a trustee to that of an owner. Further, control of water related to control over land has been an indirect way in which property rights have developed around the notion of water. Under common law the principle that flowing water is common to all was maintained but it was progressively accepted that owners of land adjacent to watercourses could make reasonable use of them.[112] Even where ownership of water is in principle not allowed, different exceptions have been allowed. The case of Islamic law is noteworthy. Original principles evolved in a highly water scarce area emphasize that water is a gift from God, that everyone has a right to water and that nobody can own it.[113] Yet, over time, the interpretation of the strict original principles has sometimes been permissive.[114] Additionally, some Islamic law scholars argue that individuals or groups have the right to use, sell, and recover value-added costs of most categories of water.[115]

110 *Justinian's Institutes* (Trans. P Birks & G McLeod, London: Duckworth, 1987) 55.

111 Constitución política de la República Oriental del Uruguay, art 47(2).

112 S Hodgson, *Land and Water—The Rights Interface* (Rome: FAO, FAO Legislative Study 84, 2004) 49.

113 NI Faruqui, 'Islam and Water Management: Overview and Principles' in NI Faruqui, AK Biswas & MJ Bino (eds), *Water Management in Islam* (Tokyo: United Nations University Press, 2001) 1, 12.

114 DA Caponera, *National and International Water Law and Administration—Selected Writings* (The Hague: Kluwer Law International, 2003) 74.

115 Faruqui (n 113 above) 12.

Thirdly, the scope of water law has only increased slowly with time. Thus, in India, the first formal water laws introduced by the colonial administration concerned, for instance, the harnessing of water for productive activities, such as irrigation, navigation and embankments.[116] At the same time, while drinking water has always been an important concern and while the special nature of water has been upheld at least in theory, law gave for a long time scant attention to drinking water issues. This is illustrated by the fact that the drafters of the Indian Constitution, like the drafters of the international covenants on human rights, failed to include an explicit mention of water in their list of fundamental and human rights.

The disjointed and sectoral nature of water law is of primary importance in the context of ongoing water law reforms. Indeed, one of the first things that need to be done is to ensure that the unitary nature of water is effectively reflected in law. This need for a coordinated and comprehensive approach is not specific to water. In fact, the same issue arises in the context of environmental issues where addressing problems in isolation does not work. Unlike water law, environmental law has evolved relatively fast towards a more comprehensive outlook. In the case of water, it is different bodies of rules and regulations developed over many decades that need to be put together and brought under a new framework. This is the major challenge that most countries of the South face. Yet, as later chapters show, current reforms do not succeed in achieving this because the agenda they set for themselves is conceptually too narrow.

C. Water Law for Poverty Eradication and Development in India

Traditional water law has provided the basis for a number of positive outcomes. Thus, in the context of the focus on harnessing water for productive activities, existing irrigation legislation provided the state with the necessary tools to foster better availability of irrigation water thereby contributing indirectly to a significant reduction in malnutrition. However, traditional water law is not appropriately equipped to deal with the challenges that India faces today and does not sufficiently reflect some significant broader changes that have occurred in the past few decades. This is, for instance, the case of irrigation acts, several of which were adopted before independence while the ones adopted since independence tend to follow the same conceptual model. Additionally, traditional water law was largely devoid of a social perspective. This has progressively changed over time with an increasing focus on drinking water in water instruments but the lack of any legislative instrument dealing specifically with drinking water despite the higher priority it is given in all policy instruments or the lack of instruments specifically implementing the human right to water are testimony to the need for significant changes.

[116] eg Inland Vessels Act 1917.

While formal water law was on the whole based on dated concepts and principles until the beginning of this decade, a number of changes have been initiated over time. These include significant joint efforts by the central and state governments to foster access to water in rural areas, in particular in the context of the Accelerated Rural Water Supply Programme.[117] Additionally, the courts took a keen interest in water during the 1980s and 1990s and started the process of updating water law by, for instance, confirming the existence of the human right to water in Indian law and formally recognizing water as falling under the notion of public trust.[118]

The changes undertaken in policy instruments or through judicial intervention notwithstanding, formal water law was in need of significant changes by the end of last century. This includes the need to update existing water laws such as dated irrigation acts, as well as the need to fill the gaps such as in the case of drinking water. Besides, Indian water law lacks a framework water legislation that sets the basic principles applicable throughout the water sector. Various countries have adopted framework water laws in recent years to address the various new challenges that they face.[119] Indeed, there is a need to move beyond the sectoral development of water law, which has been a hallmark of its development in India as well as a number of other countries for many years.

Reforms have in fact been introduced over the past decade. This is welcome in view of the general need for changes and additions to existing water law. The specific conditions under which water law reforms have been introduced calls, however, for further remarks. Firstly, it is not water law reforms that have been introduced but changes to the legal framework that are part of a broader package of measures known as 'water sector reforms'.[120] In other words, the introduction of water sector reforms preceded the introduction of a comprehensive set of changes to water law. Secondly, water law reforms that have been introduced are not based on a new set of legal principles. This is due to the fact that law reforms are conceived as a part of water sector reforms and the principles defined as a basis for the latter become the principles that underlie the development of new legal frameworks. The consequence is that law reforms that are introduced do not consider the basic principles of water law and superimpose economic principles. As a result, they are in certain cases narrowly conceived, as in the case of water user association legislation that only addresses a limited set of issues in the irrigation sector. They also mirror a specific economic logic, which is reflected in the setting up of independent water regulatory authorities meant to partly divest the state from its functions in the water sector. This also translates into a limited

[117] ch 5.A.3.
[118] ch 2.B.2.
[119] eg South Africa, National Water Act 1998 and Brazil, Law No. 9,433 of 8 January 1997 (National Water Resources Policy).
[120] ch 3.A–B.

understanding of water law reforms, which are sometimes largely equated with reforms linked to commercialization and privatization.[121]

Most of these reforms are surprising for several reasons. Firstly, the changes being introduced do not reflect recent principles recognized in the case law. Secondly, the new principles have not been debated in any legislature in India because the principles for water sector reforms are taken as a given. As a result, there have been no broad-based debates on the kind of water sector reforms and water law reforms that are required in Parliament or state legislatures. Thirdly, a number of the new laws that are introduced are directly linked to development aid conditionality. There is thus a direct link to an international policy agenda that democratically elected representatives at the local, state or national level in India have not contributed to develop. There are thus questions of process that arise even if ongoing results are deemed the most suitable choice for India at this juncture and for the next few decades.

The limitations of ongoing water law reforms call for further thinking on the type of reforms which are necessary in water law in India today. Firstly, the reforms must be based and related to existing laws where they exist. Thus, in the case of irrigation, ongoing reforms which focus on participatory irrigation management are insufficient because they only address a small number of issues in the irrigation sector while it is the whole irrigation legislative framework which is outdated and needs to be updated.[122]

Secondly, the context for water law reforms must be broader than what water sector reforms provide. This is due to the fact that water law must integrate all dimensions of water, prioritize water uses, reflect development in other areas of law such as environmental law and contribute to the realization of the human right to water and other human rights dependent on water. A broader framework for water law puts it squarely at the centre of a broadly conceived process of development. There is thus a direct link between water law and poverty eradication as well as environmental conservation, sustainability, and equity.

In policy terms, this is currently addressed largely through the MDGs highlighted above. These are indeed useful pointers to measure progress towards poverty eradication. However, from a conceptual point of view, they do not provide the basis for water laws that effectively address poverty. Indeed, the very fact of introducing targets in the form of percentages indirectly negates the idea of poverty eradication or, from another perspective, the realization of the human right to water for everyone. Only the latter, namely an absolute standard that can be qualified as being successful when each and every person has access to sufficient

[121] World Bank, Efficient, Sustainable Service for All? An OED Review of the World Bank's Assistance to Water Supply and Sanitation (Report No. 26443, 2003) 13-5.

[122] National Commission for Integrated Water Resource Development Plan, Report (New Delhi: Ministry of Water Resources, 1999) 223, acknowledges, for instance, the need for updating old irrigation acts even though it only specifically calls for participation in irrigation management.

water of appropriate quality can provide the basis for water laws that effectively address the challenges of the twenty-first century.

Most people would probably agree that water law needs to be significantly changed. However, consensus only extends to the need to initiate reforms but not on the direction that reforms must take. There is thus a need for a broader enquiry into ongoing water law reforms based on water sector reforms. While all panchayats, all Indian municipal councils, each Indian state and India as a country may decide that the types of reforms being introduced are the right reforms from the local to the national level, this can only be established if the choices are made following a process which provides all the information to all users of water—in other words everyone—and is undertaken according to all the democratic procedures that distinguish India from many other countries. This has not happened yet and there is thus a need to enquire further into the nature of existing and 'reformed' water law, its relevance to poverty eradication and the realization of human rights in theory and in practice so that the long-term choices that need to be made by democratically elected bodies from the local to the national levels are made with full knowledge of all the consequences of the choices made. This also implies that there should be no more reforms of water laws linked to any development aid conditionality since this distorts—or in the worst case scenario vitiates—democratic processes put in place under the Constitution. This book seeks to contribute to a broader debate on the types of water law reforms that are necessary and the principles of water law that are needed in the twenty-first century. While most remarks are India-specific, a number of common trends can be observed in various other countries. The findings in this book thus provide significant lessons for the further development of water law in other countries as well.

2

Evolution of Water Law

In historical perspective, water law has both been of central importance to lawyers and policy-makers and a relatively marginal area of law. This hybrid status is due to different reasons. Firstly, there has been no framework water law in India in the modern era. The consequence is that water was for a long time largely considered separately according to its different uses and its physical location. Irrigation, drinking water or groundwater were thus addressed in part as separate issues with different legal instruments and principles governing their use and different institutions taking the lead.

Secondly, a large part of water law was not codified. Thus, while irrigation acts have been progressively adopted since the late nineteenth century, rights of access and control over flowing surface water and groundwater were and remain largely governed by common law principles. Additionally, besides formal laws and norms that apply in principle everywhere, water has been governed by a multiplicity of arrangements at the local level. These customary or religious norms that are not necessarily always in conformity with the formal legal framework have often survived long after the adoption of laws displacing them because the latter were not necessarily enforced in each and every village.

Thirdly, water law has addressed since the nineteenth century a number of different concerns including navigation, construction and maintenance of embankments, fisheries, and irrigation. This variety masks the relatively greater importance accorded to access and control over water for economic development. The emphasis on water as an input for economic growth has had the concomitant impact of giving prominence to issues of control over water and the types of property rights allowed. While the principle has been that water could not be appropriated, different forms of control linked to land ownership have been condoned. As a result, issues of control over water have often been considered in the context of issues related to land more generally.

Fourthly, the emphasis put on the regulation of water use for economic growth has had the impact of restricting the scope of water law. In fact, water has often been conceived like other natural resources that need to be harnessed for economic development, for instance, through the building of dams for energy generation or irrigation. Colonial authorities tended to adhere to this model. Despite a number of changes after independence, the conceptual framework remained largely unchanged.[1] As a result, water law failed for a long time to address some of the issues that are at the centre of a modern water law regime like the human right to water.

On the whole, water has always been a central concern of policy-makers and lawyers because of the link between water and life. This explains, for instance, why there has in principle been reluctance to condone ownership claims over water. Yet, this did not translate into a water law framework that specifically and entirely reflected this basic importance of water.

The relatively narrow basis of water law has been increasingly challenged in the past few decades. This is due to many factors ranging from increasing water scarcity to water pollution affecting the quality of drinking water and the environment. A number of changes have taken place. Some of the most important include the formal recognition of the human right to water and the need to consider environmental factors as an integral part of water law. This needs to evolve further as indicated in later chapters but the very recognition that the human right to water is at the centre of water law dramatically expands the scope of water law in general.

Ongoing changes do not all pull in the same direction. Thus, at the same time as the human right to water is being given a more prominent role, the introduction of economic principles in water law has the potential to pull water law in a different direction. The emphasis on the need to conceive of water as an economic good that can be the subject of usufructuary rights or tradable rights shifts the debate back to a focus on water as an economic resource. While water law is being pulled in different directions, this leads in certain cases to common positions based on different grounds. Thus, both a human right perspective and an economic good perspective on water call for severing the link between land rights and access to water.[2]

This chapter focuses on some of the basic features of water law as they developed over time until the late twentieth century. This provides the basis for analysing in later chapters the reforms that have been introduced over the past couple of decades—and more particularly in the present decade—and that propose to completely restructure water law. Section A provides an overview of the structure of water regulation in India. Section B then examines in more detail issues related to access and control over water. Finally, Section C examines human rights and environmental aspects of water law, which are still relatively novel additions and whose effective implementation will require further measures in the future.

[1] eg C Singh, *Water Rights and Principles of Water Resources Management* (Bombay: Tripathi, 1991) 5.

[2] For more developments on this point, ch 6.B.2.

A. Water Law Framework

Water regulation has been important from time immemorial. The laws of Manu addressed issues related to the regulation of water, such as water pollution and its impacts on health.[3] In the Kautilian period, the Arthashastra specifically indicated that users had to pay a water tax for the use of water for cultivation even if the water was taken from rivers, lakes or springs.[4] Private ownership of reservoirs, embankments and tanks was allowed.[5] Further, private owners were allowed to give waters to other parties through irrigation works in exchange for a share of the produce.[6]

Similarly, Islamic law applied during the period of Muslim rulers also addressed water regulation. This recognized that water is a common resource and that everyone is entitled to free access and use of water.[7] Yet, until the colonial period it seems that there was little emphasis on formal water law in large part because water was not generally perceived as scarce. Indeed, while Islamic law had from the outset strict principles of water law developed for the water scarce region where the first Muslims lived, the need to apply these rules was never really felt in India because conditions of such scarcity did not exist at the time.[8]

The colonial government started taking a direct interest in water law in the nineteenth century. This included several kinds of interventions including laws for the protection and maintenance of embankments, the regulation of canals for navigation purposes aimed at improving these canals and levying taxes on users, the regulation of ferries as well as fisheries.[9] The colonial government gave increasing attention to irrigation in consonance with the desire to harness water for irrigation. This included the Northern India Canal and Drainage Act 1873 for large-scale irrigation works, an act that survives today in some parts of both Pakistan and India, and the United Provinces Minor Irrigation Works Act 1920 for smaller irrigation works. On the whole, colonial laws tended to focus on the economically productive uses of water and did not concern themselves either with environmental considerations or with the social aspects of water.[10]

Over time, the colonial government took an increasingly assertive position with regard to control over water, culminating with an assertion of absolute rights by

[3] *The Laws of Manu* (translated by Wendy Doniger, Harmondsworth: Penguin Books, 1991) s 4 (46).

[4] Kautilya, *The Arthashastra* (LN Rangarajan (ed), New Delhi: Penguin, 1987) s 2(24)(18).

[5] ibid s 3(9)(34).

[6] ibid s 3(9)(35).

[7] IA Siddiqui, 'History of Water Laws in India' in C Singh (ed), *Water Law in India* (New Delhi: Indian Law Institute, 1992) 289.

[8] ibid 295.

[9] Bengal Embankment Act 1882, Northern India Ferries Act 1878 and Indian Fisheries Act 1897.

[10] eg U Ramanathan, Legislating for Water: The Indian Context (Paper presented at the 3rd Common Property Conference, Washington, DC, 1992), available at <http://www.ielrc.org/content/w9201.pdf>.

the time of the enactment of the Madhya Pradesh Irrigation Act 1931.[11] The assertion of rights of control by the government over water formed part of a broader trend that saw water law focusing on regulating access to and control over water. Thus, beyond the assertion of the state's overall control over water, irrigation acts were largely concerned with the allocation of irrigation water among landowners, thus creating a direct link between real property rights and water. This indirectly led to private control over what was otherwise considered a common resource.[12] In other words, water law was for a long time concerned to a large extent with government and private control and access to water. The narrow focus of water law until the 1970s is, for instance, confirmed by the fact that the Easements Act gave landowners a limited right to pollute.[13]

Over the past forty years, there have been momentous changes in water law. Firstly, pollution of water was taken up in earnest in the 1970s.[14] Secondly, the social and human dimensions of water have become increasingly central to water law. This is visible in different contexts, from the adoption of policy measures to ensure access to drinking water by each individual to the recognition of a fundamental right to water. Thirdly, the limitations of a legal regime leaving groundwater management to landowners have been recognized even though little effective regulatory action was taken until recently. Fourthly, as part of a broader set of economic and financial reforms, an additional set of water sector reforms and water law reforms have been introduced over the past decade. These recent developments are explored in more detail in later sections of this book.

1. Basic structure of water law

Water law in India has been significantly influenced by a decision taken before independence to devolve irrigation matters to states. Thus, the Government of India Act 1935 specifically gave power to provinces concerning water supply, irrigation, canals, drainage and embankments, water storage, and hydropower. Conflicts between provinces/princely states were subjected to the jurisdiction of the Governor General who could appoint a commission to investigate the conflict if it was found to be of sufficient importance.[15]

After independence, the Constitution retained the basic scheme chosen in 1935 and gave the states a leading role in water regulation. Water was thus included in the state list in recognition of the fact that different water issues arise in different parts of the country.[16] Yet, this does not imply that the Union has no role to play in water. Firstly, with regard to the adjudication of inter-state water

[11] C Singh, Damming the Law in Narmada (on file with the author, undated) 13.
[12] Singh (n 1 above) 34.
[13] Easements Act 1882, s 28(d).
[14] Water (Prevention and Control of Pollution) Act 1974.
[15] Government of India Act 1935, ss 130 to 134.
[16] Constitution, Schedule 7, List II.

disputes, even though no agreement could be found on a specific mechanism at the time of the adoption of the Constitution, Article 262 allowed Parliament to legislate on this issue. This led to the adoption of the Inter-State River Water Disputes Act 1956. Secondly, certain powers were reserved in the 7th Schedule for the Union. The regulation of inter-state rivers was one such item which led Parliament to enact the River Boards Act 1956. This Act has, however, proved only partially successful since there is still only one broad-based river authority and a few others, such as the Narmada Control Authority, that are only involved at the operational level.[17] Thirdly, the Union has taken action in the context of article 252 of the Constitution that allows Parliament to adopt an act in a field where states are competent provided the states have given their assent. This was the basis for the adoption of the Water Act 1974. Fourthly, the Union has used less formal mechanisms to prod states into adopting certain measures. Thus, in view of the lack of progress in the provision of drinking water in rural areas, the Union came up in the early 1970s with a set of guidelines for rural drinking water supply.[18] This was never 'imposed' on states but the guidelines have been implemented and mainstreamed through the provision of finance by the Union government for drinking water schemes in states. The Union has often used financial resources as the stick that ensures states follow the policies wanted by the Union. Thus, both in the early 1970s and earlier this decade, the Union fully funded for a time specific drinking water programmes to ensure states would adopt the new policy framework embodied in the funding.[19] Fifthly, the Union has other water-related powers that it can exercise, for instance, in the context of the impact assessment of large projects that require an environmental clearance.[20]

The constitutional division of powers between the Union and states constitutes the basic framework for formal water law. However, over time, an additional important component has been added with the adoption of the 73rd and 74th amendments to the Constitution that significantly strengthened democratic governance at the local level. In rural areas, panchayat institutions are now given specific powers in the water context. This includes powers and responsibilities over drinking water supply, minor irrigation, water management and watershed development as well as fisheries.[21] Similarly, municipalities have been given powers over water supply for domestic, industrial and commercial purposes.[22] The momentous changes proposed in the constitutional amendments are yet to

[17] AD Mohile, 'Government Policies and Programmes' in J Briscoe & RPS Malik, *Handbook of Water Resources in India—Development, Management and Strategies* (New Delhi: The World Bank and Oxford University Press, 2007) 10.

[18] ch 5.A.3.

[19] In the 1970s, the Accelerated Rural Water Supply Programme Guidelines were first implemented with 100% funding from the Union. Similarly, the Swajaldhara Guidelines were implemented from 2003 with 100% funding from the Union. On the Swajaldhara Guidelines, ch 5.B.2.

[20] Government of India, The Environmental Impact Assessment Notification 2006.

[21] Constitution, art 243G and Eleventh Schedule.

[22] Constitution art 243W and Twelfth Schedule.

be fully implemented at the local level. Yet, they constitute some of the most significant changes brought to the Constitution because they have the capacity to effectively recast the distribution of power in favour of democratically elected local bodies. The importance of the democratic nature of these bodies and the constitutional sanction they have cannot be emphasized enough in the context of water. Indeed, water sector reforms make frequent reference to the democratic nature of decentralization as conceived under the constitutional scheme as a justification for the decentralization and participation agenda they put forward. As chapters, 3, 4 and 5 demonstrate in turn, there are significant differences between the two types of decentralization even though policy and academic discourses tend to assume that they are closely related.

The Constitution is also important in other ways in the water sector. Firstly, it is through the existing fundamental right to life that judges have read the existence of a fundamental right to water.[23] Secondly, a number of fundamental rights have direct bearing on water regulation. Thus, the Constitution specifies, for instance, that the prohibition of discrimination which includes sex, religion, and caste discrimination extends to the use of wells and tanks maintained wholly or partly out of state funds or dedicated to the use of the general public.[24] This is significant in the context of ongoing difficulties that specific communities face in getting access to drinking water.

B. Access to and Control over Water

Access to and control over water has been a key issue in water law for a long time. This is prima facie unexpected since water has in principle not been subject to appropriation. The importance of rules concerning access and control thus indicate a deeply entrenched tension in water law.

On the one hand, there is a strongly enshrined view that water is of such vital importance to life and human survival that it cannot be owned by anyone. Additionally, the nature of water, in particular flowing water, makes it difficult to establish ownership rules since ownership is in principle attached to a specific object.[25]

On the other hand, forms of private and state control over water have often been condoned. Separate rules have developed for different bodies of water and different uses of water. Additionally, there are different rules concerning access to water-based resources, such as fish.[26] These different rules partly stem from some of the physical characteristics of water and the tools to extract it. Thus, the

[23] *Subhash Kumar v State of Bihar* AIR 1991 SC 420 (Supreme Court of India, 1991).

[24] Constitution art 15(2)(b).

[25] eg A Hildering, *International Law, Sustainable Development and Water Management* (Delft: Eburon, 2004) 98.

[26] eg Arunachal Pradesh Fisheries Act 2006.

difficulty encountered in establishing property rights rules over flowing water does not apply for a tank built on private land. Similarly, until a few decades ago, it was assumed that landowners would not be able to exhaust the water under their land and thus would not cause any damage to neighbouring landowners. The current fragmented system also partly stems from the direct and indirect link that has often been made between access to and control over water and land ownership. Where water was seen as directly linked to the land like in the case of groundwater, control by the landowner was possible as an outcome. In situations where water flows through different plots, rules developed around the fact that surface water cannot always be controlled in the same way that immovable land can.

1. Government control

The government has often asserted or attempted to assert control over natural resources, especially with a view to control the economic benefits derived from their exploitation. Water has been no exception and the link between government interest in control and economic activities explains, for instance, the importance given to irrigation laws for a number of years.

Government control in the context of water raises issues because water is not a substance like other resources over which the government has asserted control. In general, the government asserts control because it is supposed to embody the broader public interest. This is in principle an appropriate justification but in practice the assertion of governmental control in India has not necessarily ensured that its benefits are fairly and equitably distributed across society.[27]

The control that the government asserts is derived from the sovereign nature of the state. Different consequences can be derived from this overall status. During the colonial period, the argument was, for instance, made that the state was the owner of all unoccupied and so-called waste land.[28] Today, the sovereign power of the state can be indirectly derived from the fact that in international law, there is a strong presumption that water cannot yet be considered a shared natural resource.[29] This seems confirmed by the fact that the draft articles of the International Law Commission on groundwater are still premised on state sovereignty.[30]

At the domestic level, the sovereign power has from time to time asserted strong rights of control. Thus, in the modern era, the British colonial administration asserted in some cases full control over water and the same view has been

[27] eg U Ramanathan, 'Displacement and the Law' (1996) 31/24 *Economic & Political Weekly* 1486.

[28] eg BH Baden-Powell, *A Manual of Jurisprudence for Forest Officers* (Calcutta: Superintendent of Government Printing, 1882) 88.

[29] PW Birnie & AE Boyle, *International Law and the Environment* (Oxford: Oxford University Press, 2002) 301.

[30] Draft Articles on the Law of Transboundary Aquifers, in Report of the International Law Commission, Sixtieth session, UN Doc. A/63/10 (2008) art 3.

maintained in some post-independence state laws. The power of the state to control has been challenged in theory but this has not had significant impacts in practice. This is problematic because the state has not shown itself capable of conserving, sustainably using, and equitably sharing the benefits of the exploitation of natural resources or water. In the case of water whose special nature has generally been accepted, this is even more inappropriate. Yet, it is also this special status of water which partly explains the ease with which the sovereign power has been able to assert direct and indirect control. Indeed, the general prohibition of ownership of water gave the state from the outset a relatively wide margin of appreciation in deciding how to use water for the greater common good. This progressively led to a situation where the state arrogated itself the right to use water even though this could be argued to go against the existing legal position in India.[31]

In addition, under the power of eminent domain, the state attributes itself the power to take away for a public purpose any resource under its jurisdiction.[32] In practice, eminent domain has often become an instrument of the state that is used to disempower people and redistribute rights and benefits. One of the uses of eminent domain is where the state takes over the property rights of people, for instance, their home to allow the construction of a big dam that provides benefits in the form of irrigation water, drinking water or hydropower to people in other parts of the country. These beneficiaries are often already better-off than the people affected by the use of the power of eminent domain.[33]

In legislation, the assertion of the state power can be identified in various contexts. At the federal level, the Cantonments Act specifically provides that water, which is declared a source of public water supply can be placed directly under the control of the Board, whether the source is found inside or outside of the cantonment area. Interestingly, the provision of the Colonial Act of 1924 has been retained in the new act of 2006.[34] At the state level, several irrigation acts assert complete governmental control over water. The Madhya Pradesh Irrigation Act, 1931 seems to be the earliest such assertion and indicates in clear terms that '[a]ll rights in the water of any river, natural stream or natural drainage channel, natural lake or other natural collection of water shall vest in the Government'.[35] The principles of the colonial legislation were confirmed in the Madhya Pradesh Regulation of Waters Act 1949.[36] Even the Bihar Irrigation Act adopted as late as 1997 restates word for word the provision of the colonial act.[37] The government has also sometimes claimed control over land linked to water use as in the case of

[31] eg Singh (n 11 above) 18.

[32] U Ramanathan, 'A Word on Eminent Domain' in L Mehta (ed), *Displaced by Development – Confronting Marginalisation and Gender Injustice* (New Delhi: Sage, 2008) 133.

[33] ibid.

[34] See respectively Cantonments Act 1924, s 218 and Cantonments Act 2006, s 189(1).

[35] Madhya Pradesh Irrigation Act 1931, s 26.

[36] Madhya Pradesh Regulation of Waters Act 1949, s 3.

[37] Bihar Irrigation Act 1997, s 1.

canals, irrigation works and land forming part thereof.[38] Recent debates are still informed by the desire of the state to assert control. Thus, in the context of the drive to increase the irrigation potential and hydropower generation, the Union government is planning to adopt some rivers as national assets. According to the Water Resources Minister, this would not imply 'nationalizing' rivers.[39]

The case law also provides a continued reassertion of the power of the controlling interest of the state over water. A relatively early decision of 1936 specifically indicated that the state had the sovereign right to regulate the supply of water in public streams.[40] Much more recently, the Supreme Court has restated that 'undoubtedly the state is the sovereign dominant owner' of water.[41] In the latter decision, judges understand the power of the state as extending even where there are acknowledged customary norms that govern control over water. The controlling interests of the state extend beyond public streams like a river. When a dam is erected, riparians lose their access rights as a consequence of the dam-induced submergence. Additionally, as in the case of the Sardar Sarovar dam, the state gets the exclusive right to exploit resources found within the reservoir, as well as the exclusive right over boating and water transportation.[42]

Overall, the state has shown a clear inclination to assert control over water. In certain cases, the power of the state is asserted with little restraint, as in the case of the Kumaon and Garhwal Water (Collection, Retention and Distribution) Act 1975 that sought to extinguish all customary rights in water and transfer the same to the state government.[43] Yet, overall the government has shied away from asserting control over all water in all contexts. The state often finds good reasons to assert control over water for activities having a direct bearing on economic development, like the construction of large dams for power generation or irrigation. At the same time, the assertion of power and control necessarily implies that the state takes it upon itself to provide water where it is required. In particular, any assertion of sovereign power implies that the state also acknowledges a duty to provide drinking water to all individuals. The enthusiasm for asserting control is not necessarily matched by the same eagerness to accept duties. As a result, until recently, the lack of clarity regarding ownership of water has ensured that the duties of the state are not always clearly articulated in law. This partly explains how different control regimes have subsisted over time for different bodies of water, even though the unitary nature of water does not provide a basis for such arrangements.

[38] Canals Act 1864, s 3 and Bihar Irrigation Act 1997, s 4.

[39] G Parsai, 'Centre to Adopt some Rivers as "National Assets"', *The Hindu* (31 January 2008), available at <http://www.thehindu.com/2008/01/31/stories/2008013159921200.htm>.

[40] *Secretary of State v PS Nageswara Iyer* AIR 1936 Mad 923 (Madras High Court, 1936).

[41] *Tekaba AO v Sakumeren AO* (2004) 5 SCC 672 (Supreme Court of India, 2004).

[42] Narmada Water Disputes Tribunal, Final Order and Decision of the Tribunal, 12 December 1979, s 11(V)(8).

[43] Kumaon and Garhwal Water (Collection, Retention and Distribution) Act 1975, ss 3–4.

2. Public trust

The assertion of absolute sovereign power by the state over water has never gone completely unchallenged. In recent decades, decreasing water per capita availability, physical, and social water scarcity and the progressive acknowledgement of links between water and the environment have contributed to the development of alternative understandings of water.

The notion of public trust is the concept that has been used in a number of countries to capture the need for an alternative understanding of water. In general, the notion of public trust is well developed in common law jurisdictions and has gained currency at the international level.[44] This refers to the idea that a group, a state or the international community holds certain resources in trust for the public because they are intrinsically valuable to the public and cannot be owned by any person.[45] It also implies that the trustee has a fiduciary duty of care and responsibility to the general public.

The notion of public trust has several important characteristics in the context of water. Firstly, it provides a basis for considering water without starting from the perspective of property rights. This is significant in view of the importance that land rights have acquired over time as a determinant of control over and access to water in many countries. Secondly, it provides a basis for fostering distributive justice in the sharing of and access to water. Under the public trust doctrine, the trustee is bound to distribute existing water so that it neither deprives any individual or group from access to domestic water nor significantly affects ecosystem needs.[46]

One of the consequences of the application of the notion of public trust is that the trustee can at most hold a usufructuary right in water which is deemed to be granted with the consent of the people. This also implies that where the trustee acquires usufructuary rights for public use it needs to compensate original beneficiaries which is in principle the public at large. At the very least, the trustee cannot alienate the trust nor can it fundamentally change its nature.[47] In the United States, the public trust doctrine has, for instance, been interpreted as implying that the state has a duty to protect streams, lakes, marshlands, and tidelands. The Supreme Court of California thus specifically indicated that the public trust was more than an affirmation of state power to use public property for public purposes but also a duty of the state to protect the people's common heritage, a duty that it can only surrender in rare cases and where this is consistent with the purposes of the trust.[48] One of the implications of public trust status is

[44] *Gann v Free Fishers of Whitstable* (1865) 11 ER 1305, HL.

[45] M Moench, 'Approaches to Groundwater Management: To Control or Enable?' (1994) 29/39 *Economic & Political Weekly* A135.

[46] Singh (n 1 above) 76.

[47] eg Moench (n 45 above) A140.

[48] *National Audubon Society v Department of Water and Power of the City of Los Angeles* 33 Cal 3d 419, 441 (Supreme Court of California, 1983).

that the rights of riparian landowners are subsidiary to the interest of the public to maintain rivers 'substantially undiminished, except by such drafts upon them as the guardian of the public welfare may permit'.[49]

In India, the Supreme Court has determined that flowing water is a public trust.[50] This happened in the context of a case where a club was built on encroached forest land that was subsequently submerged by the changing course of a river. The club owners were attempting to steer the river back to its old course when the case was decided. This provided the judges a basis for a discussion of the public trust doctrine. Building on US case law, they stated that

> [t]he State is the trustee of all natural resources which are by nature meant for public use and enjoyment. Public at large is the beneficiary of the sea-shore, running waters, airs, forests and ecologically fragile lands. The State as a trustee is under a legal duty to protect the natural resources. These resources meant for public use cannot be converted into private ownership.[51]

The judges specifically indicated that the conflict between conservation and economic development should preferably be taken up by the legislature. However, in view of the absence of legislation, the government acting under the doctrine of public trust cannot abdicate the natural resources and convert them into private ownership or for commercial use. In other words, the non-use value of natural resources and the environment cannot be eroded for private, commercial or other uses unless courts find it necessary in the interest of the public to encroach upon these resources.[52]

At this juncture, the public trust is clearly effective in India. Yet, the existing legal framework does not fully incorporate the principle as developed by the Supreme Court. Further, the scope of the application of the public trust remains limited since it does not apply to all waters even though there are strong arguments against a selected application that, for instance, does not include groundwater.

The fact that the notion of public trust applies at all in India is an important step forward in the legal recognition of the special nature of water. Yet, the concept of public trust is not in and of itself a solution to the problems identified with the complete assertion of power under the doctrine of eminent domain. Indeed, the core of the notion of public trust is the state's authority to control water.[53] This does not give any indication of the ways in which the state has to exercise its control. This is problematic because the main guiding factor is the notion of 'public interest'. In the case of water in the twenty-first

[49] *Hudson County Water Co v McCarter* 209 US 349, 356 (1908).
[50] *MC Mehta v Kamal Nath* (1997) 1 SCC 388 (Supreme Court of India, 1997).
[51] ibid para 34.
[52] ibid para 35.
[53] *National Audubon Society v Department of Water and Power of the City of Los Angeles* 33 Cal 3d 419, 425 (Supreme Court of California, 1983).

century this implies that the state has to foster conservation and sustainable use of water. Yet, the notion of sustainability is so vague that it does not provide a clear policy guide. The main issue is that the notion of public trust has no direct links with social concerns or human rights. There is thus no guarantee that the notion of public interest used by the Indian state would lead to the realization of human rights.[54]

The lack of a framework guiding the application of the notion of public trust proves to be problematic in India. Firstly, before the colonial state asserted control over water, early legislation such as the Northern India Canal and Drainage Act 1873 provided public trust-like powers for the state who was 'entitled to use and control for public purposes the water of all rivers'.[55] This provided the basis for the strengthening of control towards the assertion of eminent domain. Secondly, neither the early assertion of the right of the state to control uses of water nor the more recent recognition of the principle by the Supreme Court seem to provide effective safeguards against the abuse of the principle by the state. This is, for instance, illustrated by the case of a johad in Rajasthan built by villagers of Lava ka Baas village in Alwar district. The state government objected to the johad and declared that it was illegal under the Rajasthan Irrigation and Drainage Act 1954.[56] Interestingly, the Irrigation Act includes the same provision as the 1873 act which gives the government the power to determine the uses to which water should be put but does not specifically provide that the government owns the water.[57] This did not stop the government from arguing that every rain drop belongs to the Irrigation Department.[58]

The assertion of public trust status for water in India is reflected in other jurisdictions. Post-apartheid South Africa has, for instance, significantly raised the status of the notion of public trust in its new Water Act that specifies the national government's role as a trustee of the nation's water resources.[59] The minister is consequently responsible for ensuring that water is protected, used, developed, conserved, managed and controlled in a sustainable and equitable manner, for everyone's benefit. The Water Act is noteworthy for putting the social and environmental nature of water side by side. Thus, it provides that the minister is responsible for ensuring that water is allocated equitably and used in the public interest, 'while promoting environmental values'.[60] A clear priority is

[54] On the distinction between the public trust and fundamental human rights, D Takacs, 'The Public Trust Doctrine, Environmental Human Rights, and the Future of Private Property' (2008) 16 *New York University Environmental L J* 711.

[55] Northern India Canal and Drainage Act 1873, as applicable to State of Uttar Pradesh, preamble. Also Bengal Irrigation Act 1876.

[56] For an account of this case, eg Centre for Science and Environment, Alwar Dam Update (2001) available at <http://www.cseindia.org/html/extra/dam/index_story.htm>.

[57] Rajasthan Irrigation and Drainage Act 1954, s 5.

[58] N Jha, 'Traditional Minor Irrigation Mechanisms: State Versus Community Conflicts' (2004) 1 *Indian Juridical Rev* 244.

[59] South Africa, National Water Act 1998, s 3(1).

[60] ibid s 3(2).

given to social aspects of water but the act also acknowledges that social priorities are influenced by environmental aspects. This is an important statement because of the increasing tendency to use environmental conservation ahead of social considerations in particular where the balance is between the poorest and 'beautification'. The South African example indicates that, with appropriate guidelines, the public trust can be relevant today even though its successful application in practice hinges on government restraint in the use of its powers as trustee.

3. Individual access to and control over surface water

In principle, flowing water is either under the sovereign control of the state or covered by the notion of public trust. Yet, this has not stopped the development of a number of individual entitlements such as rights of access to water, rights to use water or rights to use water-based resources such as fish. Most of these entitlements are usufructuary in nature.[61]

In certain cases, individuals have been granted entitlements to appropriate water either on a first-come first-served basis or by virtue of their proximity to a water source. In this case, the water flowing through a river cannot be owned but individuals can appropriate it for their own private uses. The theory of riparian rights was adopted in common law jurisdictions. It is based on a preference given to the claims of landowners over waters passing through or bordering their lands. The theory developed mostly in England and North America in the nineteenth century providing landowners with rights that were an integral part of the right of ownership.[62] It provides landowners with a usufructuary right to use a portion of the flow of a watercourse. Originally, riparianism was based on the idea that *in situ* use of the water was the norm. Over time, *ex situ* or consumptive uses have been increasingly acceptable and in some countries reasonable use which includes consumptive uses is allowed.[63]

The system of riparian rights seems to have functioned well as a basis for water distribution as long as it was applied in climates and places where water was available in plenty. Whenever it has been applied in more arid places, it has usually given way to other sharing principles given that riparian rights do not provide a basis for sharing water in a socially equitable and environmentally sustainable manner.[64]

[61] eg S Hodgson, Land and Water—The Rights Interface (Rome: FAO, FAO Legislative Study 84, 2004).

[62] ibid 49.

[63] AD Tarlock, 'Water Transfers: A Means to Achieve Sustainable Water Use' in E Brown Weiss, L Boisson de Chazournes & N Bernasconi-Osterwalder (eds), *Fresh Water and International Economic Law* (Oxford: Oxford University Press, 2005) 35.

[64] cf Singh (n 1 above) 69.

Despite the fact that a large part of the world has been facing conditions of water stress for a long time, riparian rights were incorporated in direct or indirect ways in a number of colonized countries. This was the case in India as well.[65] Riparian rights are, for instance, statutorily confirmed under the Limitation Act.[66] The rights of riparian owners are not absolute but limitations largely relate to the interests of other owners and their own right to the quantity of water they are customarily entitled to draw.[67] The restrictive nature of riparian rights linked to property rights concerns has increasingly been challenged. Thus, in South Africa where unequal land right distribution and apartheid were closely linked, riparianism was completely abolished under the post-apartheid Water Act 1998.

The limitations of the original theory of riparian rights led to the development of the doctrine of prior appropriation. This was originally proposed in the arid areas of western United States as an alternative to riparianism. Under prior appropriation, the first user of water that puts it to beneficial use has priority and his/her claim takes precedence over subsequent users.[68] This doctrine therefore privileges whoever first starts putting water to a use determined as being beneficial, ranging from agricultural and economic to domestic uses and more recently ecological purposes as well. The right is not directly related to land ownership and depends on continuous application to beneficial use. The limitations of this system which privileges individual claims over the broader interests of society and of the water system have led concerned states to posit that water is not owned by individuals but by the state in trust for the public.[69]

Another type of individual entitlement concerns rights to use a specific quantity of water. This can take the form of a water licence for different uses such as irrigation.[70] This entitlement is usually linked to property rights in land or at least a limited claim to land like the one of the farmer renting a plot for farming. These rights are transferred with the land.[71] There is no right to the water itself but a right to a certain allocation of water which may be conditioned, for instance, on actual availability in a given year. These types of entitlements have been the object of much attention in recent years with the introduction of participatory irrigation management schemes. In this context, there are increasing debates over the nature of the rights that should be granted to individual land holders and occupiers. In particular, there have been proposals to delink the nexus

[65] *Secretary of State v S Subbarayudu* AIR 1932 Privy Council 46 (Privy Council, 1931).

[66] Limitation Act 1963, s 25.

[67] *Ramsewak Kazi v Ramgir Choudhury* AIR 1954 Patna 320 (Patna High Court, 1954).

[68] eg Hodgson (n 61 above) 26.

[69] Tarlock (n 63 above) 44.

[70] R Meinzen-Dick & L Nkonya, 'Understanding Legal Pluralism in Water and Land Rights—Lessons from Africa and Asia' in B Van Koppen, M Giordano & J Butterworth (eds), *Community-Based Water Law and Water Resource Management Reform in Developing Countries* (Wallingford: CABI, 2007) 12.

[71] eg Orissa Irrigation Act 1959, s 24.

between water and land with a view to facilitate trading in water entitlements separately.[72]

Finally, what is in some ways the most evolved form of water entitlement concerns situations where an individual or legal entity is given not only the right to a quantum of water but also control over the water flow, its use and resources. At present, this seems to remain an exception. However, there are situations like in the case of the river Sheonath in the state of Chhattisgarh in India where a de facto privatization was sanctioned in 1998 through the lease of a stretch of river to a company.[73]

4. Access to and control over groundwater

Groundwater has usually been treated separately from surface water.[74] Historically, this can be ascribed in part to a lack of understanding of the connections between surface and groundwater and of the relationship between groundwater abstraction in different places. This also reflected the unavailability of pumping devices allowing large-scale groundwater withdrawals to the extent of significantly affecting the water table level.

These factors contributed to the development of separate legal principles for control over and use of groundwater. Since groundwater has a direct link to the land above, a link was established between ownership of the land and control, if not outright ownership, of the water found underneath the plot. While no specific groundwater legislation arose until the past decade, basic principles of access and control can be derived from the Easements Act 1882. Under these principles, landowners have easementary rights to collect and dispose of all water found under their land.[75] There is thus an indissociable link between land ownership and control over groundwater. This implies that groundwater is mostly controlled by individuals or legal entities that own or occupy land. Where the common law principle is strictly applied, landowners are not restricted in the amount of percolating water they can appropriate.[76] It can, however, be argued today that, even under common law principles, owners cannot exploit groundwater beyond the replenishable level.[77]

The link between groundwater and land ownership is important for different reasons. Firstly, groundwater has been and is an increasingly important source of drinking water. This is due both to the existence of increasingly

[72] ch 4.B.2 concerning the Maharashtra Water Resources Regulatory Authority Act.

[73] ch 3.B.6, p. 80.

[74] This also holds in other parts of the world. For southern Africa, eg LA Swatuk, 'The New Water Architecture of SADC' in DA Mcdonald & G Ruiters (eds), *The Age of Commodity—Water Privatization in Southern Africa* (London: Earthscan, 2005) 43.

[75] *Halsbury's Laws of India—Volume 29(2)* (New Delhi: Butterworths, 2000) 447.

[76] Moench (n 45 above) A137.

[77] Ground Water Management and Ownership—Report of the Expert Group (New Delhi: Government of India, Planning Commission, 2007) 23.

powerful pumping devices as well as to an increasing bias against the use of surface water as a source of drinking water to ensure that it is of better quality. Secondly, groundwater has been an increasingly important resource used by landowners in different types of economic activities. In fact, groundwater has now become in certain regions as important or even more important than land itself.[78] Besides agriculture, large-scale water abstraction is also carried out by certain industries, as in the case of water or soft drink bottling plants.

Where control over groundwater is linked to land rights, there are neither any incentives for individual landowners to sustainably use the resource nor any way to implement policies that take into account the welfare of a broader community and the environment. In what is for all practical purposes an unregulated system, there is, for instance, no authority that can determine how many wells, handpumps, and other tubewells can be sunk in a given area. Some form of regulation that takes into account the broader aspects of groundwater use is thus necessary. Regulation is also required because the increasing use of groundwater controlled by private individuals may shift away control over water from communities. Thus, in the case of tank irrigation in Tamil Nadu that are often largely community managed, increased use of groundwater and the lesser importance attached to tanks seems to have shifted the determinants of water access away from communities into the hands of individuals.[79]

The importance of groundwater to all water users can become a source of conflict where abstraction of water on one plot of land affects water availability on neighbouring plots. This has been highlighted in recent years in the case of conflicts between bottling plants set up in areas surrounded by habitations or fields.[80] Another increasingly important issue is the quality of groundwater. Alongside the rising importance of groundwater, threats to its quality have also dramatically risen. These include direct impacts, as in the case of industries that directly pump their waste water into the ground.[81] Indirect threats include situations like the case of the Yamuna, which is so polluted after its passage through Delhi, that farmers residing dozens of kilometres downstream cannot cultivate on its flood plains, an area that would have been the most fertile of all earlier, now largely barren.

[78] S Janakarajan & M Moench, Are Wells a Potential Threat to Farmers' Wellbeing? The Case of Deteriorating Groundwater Irrigation in Tamilnadu (Chennai: MIDS, Working Paper No. 174, 2002).

[79] ibid 2.

[80] Conflicts have arisen in different parts of the country, such as Mehdi Ganj, Uttar Pradesh and Plachimada, Kerala. On the former, eg G Drew, 'From the Groundwater Up: Asserting Water Rights in India' (2008) 51 *Development* 37 and on the latter S Koonan, 'Groundwater—Legal Aspects of the Plachimada Dispute' in P Cullet, A Gowlland-Gualtieri, R Madhav & U Ramanathan (eds), *Water Law at the Crossroads—National and International Perspectives With Special Emphasis on India* (New Delhi: Cambridge University Press, 2009) 158.

[81] Concerning the city of Aligarh, eg 'TCE Cripples' (2003) 12/5 *Down to Earth*.

The dramatic increase in groundwater use and importance of groundwater as a source of water have led to significant debates but relatively little by way of concrete policy decisions. To-date, the most significant initiatives at the Union level have been the drafting of a model bill for adoption by the states and the setting up of the Central Groundwater Authority mandated to regulate and control the use of groundwater.[82] Its mandate includes the notification of 'over-exploited' and 'critical' areas and the regulation of groundwater withdrawal in such areas but it does not have a broad mandate to regulate groundwater in general. The Authority is not credited with having had much impact in its decade of existence.[83]

These developments amount to relatively little in a context where groundwater was governed for a long time by principles that assumed self-regulation. The dramatic changes that have taken place in the past few decades and turned groundwater into the major source of water are not reflected in the existing legal framework, including in the few states that have adopted the model bill as a prototype for their legislation, since this is not a comprehensive regulatory response. This can be ascribed to two major factors. Firstly, the absence of existing models for either regulating groundwater separately or for redrafting water laws to cover all water have likely hampered efforts to give groundwater the central place it deserves. Secondly, the fact that falling water tables can be 'fixed' for some time by simply digging further down has provided an opportunity for governments to avoid facing some difficult political choices. In fact, in a number of states, the answer to falling water tables has been not to address the issue itself. State governments have thus often chosen to increase power subsidies to make extraction of ever deeper layers of groundwater possible rather than tackle the underlying cause of depletion. The limits of an approach that not only refuses to control access to groundwater but seeks to encourage it with specific subsidies have been clearly understood. The unavoidability of a different response has dawned on most states but the fact that it is an extremely sensitive political issue implies that some states may still further delay necessary measures by a number of years.

5. Customary rules of access and control

As noted at the outset, formal water law has been comparatively under-developed. Yet, because water has always been so central to human life, most communities have evolved their own local rules and principles over time. Customary rules have thus played—and still play—a significant role in shaping up individual people's actual access to water at the local level. In practice, customary norms are in

[82] Ministry of Environment and Forests, Gazette Notifications SO38 and SO1024 of 14 January 1997 and 6 November 2000.

[83] eg T Shah, 'Groundwater Management and Ownership: Rejoinder' (2008) 48/17 *Economic & Political Weekly* 116, 118.

widespread use and constitute an important, though largely hidden, component of water law.

Customary rules have in many cases survived even where formal laws and regulations have been adopted. This is due in part to the fact that the latter have often not been implemented beyond certain limited areas. Additionally, customary rules of access and control are not necessarily invalid in the eyes of the law. They may, however, be trumped.[84]

In practice, customary rules are often simply ignored. This is apparent, for instance, in situations where the adoption of legislation extinguishes existing systems of access and control without referring to the changes that the new framework introduces. This is not surprising in a context where custom is often only regarded as a source of law when it is recorded in statutes or recognized by courts.[85]

Customary rules deserve greater consideration. Indeed, a number of schemes, especially those fostering common management of water sources such as tanks and ponds have proved to be long-term sustainable solutions to the problem of water availability for many communities.[86] While the trend since independence has been towards reduced interest in these common water sources, there is an increasing interest in reviving collective institutions. Such efforts must, however, be seen in the context of the widespread adoption of water user associations, a much more important trend in recent years.[87]

Customary rules that are tailored to the situation of specific villages appeal as a general proposition in the context of the search for more decentralized governance. These rules can, however, also be extremely inequitable. This is, for instance, the case of caste discrimination in access to domestic water. In this context, while discrimination in access to water has been banned since the adoption of the Constitution,[88] caste inequalities in access to water are a reality that many people still face on a daily basis.[89]

C. Human Rights and Environmental Aspects

The preceding section has highlighted some of the main principles of established water law. The focus on access and control has proved problematic because it does not provide a basis for a comprehensive understanding of the issues related to water use. Indeed, water must be considered in the broader context in which it

[84] eg *Secretary of State v PS Nageswara Iyer* AIR 1936 Mad 923 (1936).

[85] eg V Upadhyay, 'Customary Rights over Tanks' (2003) 38/44 *Economic & Political Weekly* 4643.

[86] eg D Mosse, *The Rule of Water—Statecraft, Ecology and Collective Action in South India* (New Delhi: Oxford University Press, 2003).

[87] ch 4.A.

[88] Constitution, art 15.

[89] R Tiwari, 'Explanations in Resource Inequality—Exploring Schedule Caste Position in Water Access Structure' (2006) 2/1 *Intl J Rural Management* 85.

arises. Thus, environmental or energy policies of the state have significant impacts on water availability and hence access to water at the individual or local level. Environmental aspects are, for instance, relevant because inappropriate environmental measures may contribute to the pollution of water sources used for drinking water. Similarly access to drinking water may be affected by measures such as large-scale water diversion as envisioned under the interlinking of rivers schemes.

Over the past couple of decades, there has been a broadening of the scope of water law to include issues such as the links with human rights and the environment. This has taken place alongside the development of these fields and the realization that it is, for instance, not possible to circumscribe environmental law to a limited set of concerns and values. This section examines in turn developments concerning the human right to water and links between water law and the environment. It covers the evolution that has to a large extent already taken place and is at least in principle part of the existing legal framework. Chapter 6 reflects on these issues further and suggests ways to strengthen the place of human rights and the environment in water law and to broaden the understanding of certain key concepts.

1. The human right to water

The human right to water is increasingly widely recognized at the international and national levels.[90] Under international law, the human right to water is for instance, recognized in the Convention on the Right of the Child and has been made explicit in the context of the Covenant on Economic, Social and Cultural Rights by the eponymous Committee.[91] The UN General Assembly has also confirmed the existence of a human right to water.[92]

At the national level, various countries have formally recognized the right to water.[93] Thus, the South African Constitution expressly recognizes a right to have access to sufficient water.[94] Similarly, in Uruguay, since 2004 the Constitution provides that access to potable water and access to sanitation are fundamental human rights.[95] In the European region, a right to water is implemented but it is not necessarily always justiciable because justiciability is often linked to civil and political rights.[96]

[90] eg M Fitzmaurice, 'The Human Right to Water' (2007) 18 *Fordham Environmental L Rev* 537.

[91] Convention on the Rights of the Child, New York, 20 November 1989, 1577 UNTS 3, art 24.

[92] United Nations General Assembly Resolution 54/175, The Right to Development, 17 December 1999, UN Doc. A/RES/54/175, para 12.

[93] For a list of countries and the respective provisions, see <http://www.ielrc.org/water/doc_hr.php>.

[94] South Africa, Constitution, s 27(1)(b).

[95] Constitución política de la República Oriental del Uruguay, art 47.

[96] eg H Smets, 'Le droit à l'eau, un droit pour tous en Europe' (2007) 37/2-3 *Environmental Policy & L* 223, 225 arguing that the right to water is not a human right in its usual sense because it is not a civil and political right. He nevertheless accepts that the right may also be conceived as a socio-economic right and that this is a counter-argument to the fact that it is not a fundamental right.

In India, the Constitution does not specifically include a fundamental right to water. Yet, a number of judicial pronouncements have made it clear that such a right exists under Indian law. The Supreme Court has repeatedly derived a fundamental right to water from the right to life.[97] It sees the unavailability of drinking water to all citizens as constituting a violation of UN human rights instruments and the right to life under the Constitution.[98] High courts have also taken decisions that confirm the existence of a fundamental right to water. The Kerala High Court found, for instance, that '[t]he right to sweet water, and the right to free air, are attributes of the right to life, for, these are the basic elements which sustain life itself'.[99] The High Court of Karnataka confirmed this but specifically indicated that the fundamental right only includes drinking water and not irrigation water.[100]

Under Indian law, there is thus a clear recognition of the fundamental human right to water. Yet, the actual content of the right has not been elaborated upon. Thus, it is not possible to take the analysis much further on the basis of judicial pronouncements. This is in fact not a specificity of India but is an issue that arises in different parts of the world.[101] The rest of this section thus examines the human right to water from the perspective of existing international law instruments.

At the international level, the human right to water is firmly entrenched. It has been explicitly recognized in a variety of contexts, from the Convention on the Rights of the Child to UN General Assembly resolutions and resolutions of UN human rights bodies. Similarly, at the regional level, even though the African Charter on Human and Peoples' Rights does not explicitly recognize a right to water, the African Commission has specifically indicated that a government's failure to provide safe drinking water constitutes a violation of the right to health.[102] These documents show that the recognition of the right exists both as a human right to water per se and as the recognition of a human right to water read into existing human rights such as the right to life, health or food.

The widespread recognition of the human right to water does not imply that there is unanimity around it. Indeed, some commentators discuss the human right to water in the conditional tense.[103] Others argue that the right is not clearly

[97] eg *Subhash Kumar v State of Bihar* AIR 1991 SC 420 (Supreme Court of India, 1991).

[98] *Narmada Bachao Andolan v Union of India* AIR 2000 SC 3751 (Supreme Court of India, 2000).

[99] *FK Hussain v Union of India* AIR 1990 Ker 321 (High Court of Kerala, 1990).

[100] *Venkatagiriyappa v Karnataka Electricity Board* 1999 (4) Kar LJ 482 (High Court of Karnataka, 1998).

[101] United Nations Development Programme, *Human Development Report 2006—Beyond Scarcity: Power, Poverty and the Global Water Crisis* (New York: UNDP, 2006) 9.

[102] *Free Legal Assistance Group v Zaire* Communications No. 25/89, 47/90, 56/91, 100/93 (African Commission on Human and Peoples' Rights, 1995).

[103] eg SMA Salman & S McInerney-Lankford, *The Human Right to Water—Legal and Policy Dimensions* (Washington, DC: World Bank, 2004) 66.

defined at the international law level.[104] Such critiques are partly linked to the fact that the Covenant does not expressly mention water.

Yet, as Gleick already argued more than a decade ago, if water is not considered to be part of the rights protected by the Covenants, this 'would mean that there is no right to the single most important resource necessary to satisfy the human rights more explicitly guaranteed by the world's primary human rights declarations and covenants'.[105] Indeed, the realization of the right to water is a precondition for the fulfilment of many other fundamental rights such as the rights to life, food and health.[106] The right to water is also an intrinsic part of the content of many rights such as the rights to life, food and health.[107]

Additionally, the human right to water is one of the rights that are unquestionably part of any catalogue of fundamental human rights. Indeed, water falls without doubt within the scope of natural rights and is clearly one of the 'rights held simply by virtue of being a human person [that] are part and parcel of the integrity and dignity of the human being'.[108] This is not surprising since similar positions have been advocated since antiquity. Plato thus argued that water is the most basic necessity for human beings while Aristotle gave a clear priority to water for life over other uses of water.[109]

At the international level, like in India, statements on the right to water tend to be limited to broad general pronouncements. Yet, there is one exception to this trend which is the Committee on Economic, Social and Cultural Rights' General Comment 15 on the right to water.[110] This General Comment has the advantage of not only confirming the existence of the right in the context of the Covenant but also to provide a fairly elaborate reading of its content.

At the outset, the Committee sees the right as entitling 'everyone to sufficient, safe, acceptable, physically accessible and affordable water'.[111] It adds that

[104] J Scanlon, A Cassar & N Nemes, Water as a Human Right? (Gland: IUCN, Environmental Policy and Law Paper No. 51, 2004).

[105] PH Gleick, 'The Human Right to Water' (1999) 1/5 *Water Policy* 487, 493.

[106] eg Sub-Commission on the Promotion and Protection of Human Rights, Resolution 2000/8, Promotion of the Realization of the Right to Drinking Water and Sanitation, Report of the Fifty-Second Session, Geneva, 31 July–18 August 2000, UN Doc. E/CN.4/2001/2-E/CN.4/Sub.2/2000/46, para 2.

[107] cf Sub-Commission on the Promotion and Protection of Human Rights, Resolution 2006/10, Promotion of the Realization of the Right to Drinking Water and Sanitation, UN Doc. A/HRC/2/2-A/HRC/Sub.1/58/36 (2006) para 2 highlighting the links with other human rights.

[108] R Higgins, *Problems and Process—International Law and How We Use It* (Oxford: Clarendon, 1994) 96.

[109] H Ingram, JM Whiteley & R Perry, 'The Importance of Equity and the Limits of Efficiency in Water Resources', in JM Whiteley, H Ingram & R Perry eds, *Water, Place, and Equity* (Cambridge, Mass: MIT Press, 2008) 1, 8–9.

[110] Committee on Economic, Social and Cultural Rights, General Comment 15: The Right to Water (Articles 11 and 12 of the International Covenant on Economic, Social and Cultural Rights), UN Doc. E/C.12/2002/11 (2003) [hereafter General Comment 15].

[111] ibid para 2. This definition was substantially adopted in the Berlin Rules on Water Resources, International Law Association, Report of the Seventy-first Conference—Berlin (2004).

drinking and domestic water use has priority over all over water uses required to meet core obligations under each of the rights recognized under the Covenant.[112]

Some of the terms of this definition are given further attention in the General Comment. Firstly, the Committee examines the issue of the amount of water necessary for the fulfilment of the right but does not specify a minimum amount.[113] It restricts itself to proposing that this should be examined from the perspective of the amount of water necessary to prevent death from dehydration, reduce the risk of waterborne diseases and provide for cooking, personal and hygienic needs.[114] The reference to WHO guidelines on the question of quantity of water is insufficient since these may be read by different people as requiring anything from 20 to 100 litres as a necessary minimum.[115] This can be compared with the draft General Comment, which was suggesting specific figures but chose to refer to the two lower estimates of the WHO rather the higher ones without providing a specific rationale for the same.[116] Another pointer is a report of the United Nations High Commissioner which argues that between 50 and 100 litres of water per person per day are needed to ensure that all health concerns are met and that a threshold of 25 litres per person per day represents the lowest level to maintain life. This amount raises health concerns because it is insufficient to meet basic hygiene and consumption requirements.[117]

Secondly, the General Comment specifies that safe water is water that is free from micro-organisms, chemical substances, and radiological hazards that constitute a threat to a person's health.[118] One of the core obligations of member states with regard to the right to water is in fact to prevent, treat, and control diseases linked to water, something which is directly related to the provision of safe water. The General Comment also specifies that water must be of an acceptable colour, odour, and taste for each personal or domestic use. In other words, it recognizes a cultural component to the basic right to water.

Thirdly, the General Comment addresses the issue of physical accessibility. The Committee specifies that domestic water must be accessible 'within, or in the immediate vicinity, of each household, educational institution and workplace'.[119] In view of the fact that hundreds of millions of people do not have physical access to water in the sense highlighted here, this is a very important provision. However, the content of this section is diluted in the section dealing with member

[112] General Comment 15 (n 110 above) para 6.

[113] On the issue of quantity in India, ch 5.A.

[114] General Comment 15 (n 110 above) para 2.

[115] cf G Howard & J Bartram, Domestic Water Quantity, Service Level and Health (Geneva: World Health Organisation, 2003).

[116] Committee on Economic, Social and Cultural Rights, Draft General Comment No. 15, UN Doc. E/C.12/2002/11 (July 2002).

[117] United Nations High Commissioner for Human Rights, Report on the Scope and Content of the Relevant Human Rights Obligations Related to Equitable Access to Safe Drinking Water and Sanitation under International Human Rights Instruments, UN Doc. A/HRC/6/3 (2007) 11.

[118] General Comment 15 (n 110 above) para 12(b).

[119] ibid para 12(c).

states' core obligations which restricts itself to requiring that states must ensure physical access to water facilities or services that are at a reasonable distance from the household.[120]

Lastly, the Committee addresses the issue of economic accessibility or affordability. It is the last element of the definition but is key to the understanding of the human right to water that the Committee proposes. It is also the most controversial aspect because it drastically reduces the import of the other components of the right. The general idea that underlies the concept of economic access is that states should ensure that water is affordable for everyone.[121] Ways in which states discharge of their responsibilities under this clause include technical and technological solutions that focus on low-cost options as well as socio-economic measures such as free or low-cost water and income supplements.[122] Nevertheless, nowhere does the General Comment indicate that access to the minimum amount of water which is deemed to be necessary to fulfil the criteria of the right to water has to be provided free to all or even to all poor people. Similarly, the High Commissioner for Human Rights stops at asserting that '[a]ffordability requires that direct and indirect costs related to water and sanitation should not prevent a person from accessing safe drinking water and should not compromise his or her ability to enjoy other rights'.[123] Within the notion of economic accessibility, the General Comment includes some safeguards. It provides, for instance, that payment for water services has to be based on the principle of equity, a principle that is deemed to imply that poorer households should not be disproportionately burdened with water expenses as compared to richer households. Further, it emphasizes the need to distribute equitably all available water services and facilities and to provide them on a non-discriminatory basis, especially for marginalized groups.[124]

Overall, the General Comment proposes a right to water which is partly based on a conception of water as an economic good as proposed in the context of water sector reforms. This is confirmed by the fact that the Committee does not take a position with regard to water services privatization.[125] The point is not whether the Committee is in favour of or opposed to water services privatization. Rather, the problem is that the Committee refuses to engage with what is after all one of the most sensitive issues concerning water management.[126] Privatization does

[120] ibid para 27.
[121] ibid para 26.
[122] ibid para 27.
[123] United Nations High Commissioner for Human Rights, Report on the Scope and Content of the Relevant Human Rights Obligations Related to Equitable Access to Safe Drinking Water and Sanitation under International Human Rights Instruments, UN Doc. A/HRC/6/3 (2007) 15.
[124] ibid paras 37(b) and (e).
[125] E Riedel, 'The Human Right to Water and General Comment No.15 of the CESCR' in E Riedel & P Rothen (eds), *The Human Right to Water* (Berlin: Berliner Wissenschafts-Verlag, 2006) 19.
[126] M Craven, 'Some Thoughts on the Emergent Right to Water' in E Riedel & P Rothen (eds), *The Human Right to Water* (Berlin: Berliner Wissenschafts-Verlag, 2006) 37.

matter from the point of view of the realization of the right to water as it may have detrimental consequences from a human rights point of view, for instance, where access to water is disconnected.[127]

The General Comment also makes interesting reading with regard to the question of the violation of the right. In the context of the progressive realization of the right, the Committee specifically indicates that the adoption of retrogressive measures incompatible with the core obligations of the right to water, the formal repeal or suspension of legislation necessary for the continued enjoyment of the right to water, or the adoption of legislation or policies which are manifestly incompatible with pre-existing domestic or international legal obligations in relation to the right to water constitute violations of the right.[128] In principle, no state is likely to be found defaulting on such general commitments. Yet, ongoing water sector reforms that propose the progressive disengagement of the state from the provision of drinking water and the imposition of capital cost sharing on users can constitute violations of the right.[129] Indeed, ongoing drinking water reforms in India may end up constituting a violation of the human right to water.[130]

Overall, the General Comment does not necessarily reflect the only possible conception of the human right to water or the way in which it is understood in India. Yet, it provides a framework within which the right can be analysed and further developed. Further examination of the implementation of the human right to water is taken up in chapter 5 in the context of the focus on drinking water. Chapter 6 then considers ways in which a new understanding of the right could be developed to address some of the existing shortcomings identified.

2. Sustainability and environment

The environment was for a long time either ignored or sidelined in water law. Emphasis was, for instance, put on ways to develop water infrastructure to ensure economic growth. Over the past few decades, in keeping with the rapid development of environmental law, environmental considerations have been integrated in direct and indirect ways in water instruments. This emphasis on environmental issues can be partly attributed to a greater awareness concerning environmental impacts of water policies and the importance of appropriate environmental policies for the water sector. This is also linked to the increasing focus of international policy makers on the links between economic, social, and environmental considerations. Indeed, links can be traced between the growth of the law of sustainable development and increasing concerns for environmental aspects in water law.

[127] Disconnections are analysed at ch 6.C.2.
[128] General Comment 15 (n 110 above) para 42.
[129] cf EB Bluemel, 'The Implications of Formulating a Human Right to Water' (2004) 31/4 *Ecology LQ* 957.
[130] ch 5.C.

Environmental considerations are relevant in a number of ways in the context of water. Firstly, there is a general link between water abstraction for human use and ecological functions. This is an aspect which was sidelined for decades when the mantra of water policies was that water flowing to the sea was 'wasted' water.[131] From this standpoint, the 'unused' water potential can be extremely high as most rivers of the world still flow to their deltas. In India, for instance, the figure of 'unused' water is put at 90 per cent.[132] This is a theoretical understanding which does not take into account the variety of uses of water that need to be taken into account, including ecosystem needs. This has now changed—at least in principle—and it is widely understood that there cannot be unlimited water abstraction or diversion without negative ecological consequences. One answer is the concept of environmental flows, which highlights the fact that it is not only paramount to ensure minimum flows in rivers where dams are built but also to ensure that releases are timed according to ecological cycles and not only according to the priorities of irrigation.[133] A broader answer is the notion of reserve which looks not only at the quantity of water but also its quality and recognizes the need to generally protect aquatic ecosystems in order to secure ecologically sustainable development and use of the relevant water resource. This constitutes a way to enshrine the recognition that water use needs to be governed by its impact on the environment.

Secondly, the close link between agriculture and water implies that there are a number of direct and indirect links between the environment, water, and agriculture. Surface irrigation undertaken through the diversion of river water not only causes environmental impacts on the river ecosystem but also on the ecosystems that benefit from the additional water. The 'beneficial uses' of water for irrigation can also have a number of side-effects such as waterlogging or salinity over time. Further, there are links between the types of crops grown in irrigated areas and the environment. Thus, while irrigation can be beneficial for various types of crops, it plays a more important role in the successful introduction of hybrid varieties. Yet, the higher yield potential of hybrid varieties is not unleashed only through water availability but requires other inputs such as fertilizers and pesticides. The latter have the unfortunate side-effect of contributing to polluting groundwater, unleashing a stream of long-term consequences which are becoming increasingly significant in view of the increasing importance of groundwater for both domestic uses and irrigation.

Thirdly, there are also links between water and broader environmental issues like global warming. There are now a number of consequences of global warming

[131] Report of the Narmada Water Resources Development Committee—Khosla Report (Government of India, Ministry of Irrigation and Power, 1965).

[132] A Gulati, R Meinzen-Dick & KV Raju, *Institutional Reforms in Indian Irrigation* (New Delhi: Sage, 2005) 31.

[133] eg CJ Vörösmarty, 'Fresh Water' in R Hassan, R Scholes & N Ash, *Ecosystems and Human Well-being—Current State and Trends Volume 1* (Washington: Island Press, 2005) 170.

on water that have been identified. These include reductions in water availability as rainfall declines and temperature rises in areas like the Sahel; accelerated glacial melt leading to reduction of water availability in a number of countries in East and South Asia and Latin America; disruption of monsoon patterns in South Asia; and rising sea levels resulting in freshwater losses in river delta systems in countries such as Bangladesh and Egypt.[134] These broad-ranging impacts indicate that the links between water and the environment strengthen over time. This has obvious implications from a law and policy perspective. While environmental law has relatively rapidly developed into a corpus of rules that at least clearly acknowledges a multiplicity of links between environmental issues and other concerns such as social and economic development, this has not been the case with water law. As a result, the kind of impacts that are now expected from climate change on water mean that water law will have to rapidly evolve to become much more responsive to the links with environmental issues.

Fourthly, there are also positive links between the environment and water. Judicious environmental measures can, for instance, contribute to increasing supply of clean water. An interesting example of such an approach concerns the unlikely case of New York City. In the face of increased human development, the state decided that instead of investing in a water filtration facility it would implement catchment protection strategies to regulate the input of non-point source pollutants.[135]

Fifthly, environmental and water objectives can meet in a number of cases. Thus, rainwater harvesting has clear benefits for both the environment and access to water. Rainwater harvesting provides a way to restrict stress put on existing water resources thus contributing to better overall availability for ecological or human needs. Rainwater harvesting includes not only rooftop individual harvesting but also all the ingenious ways that people have devised over time to collect rain from check dams to tanks. This is beneficial in terms of human needs as well as for the environment since it contributes additional temporary water bodies while fostering better recharge of the groundwater. This is especially important in parts of the world like South Asia that receive most of their rainfall in a short concentrated span of a few months. Rainwater harvesting has in fact become one of the key environment-related activities in India.[136]

The links between environmental considerations and water have been taken up in three different ways in recent years. Firstly, international and national water law has incorporated some environmental concerns. Secondly, environmental law has developed a number of principles which also apply in the context of water.

[134] UNDP (n 101 above) 15.

[135] Vörösmarty (n 133 above) 179.

[136] Several states have adopted rainwater harvesting regulations. eg Tamil Nadu Municipal Laws (Second Amendment) Ordinance 2003 and Kerala Municipality Building (Amendment) Rules 2004.

Thirdly, ongoing water sector reforms are largely premised on the need to address environmental aspects of the water sector.[137]

In existing water law, environmental concerns are in certain cases an important dimension of existing legal instruments. Thus, conservation and preservation of water constitute two of the main dimensions of the UN Water Convention.[138] Environmental aspects are even more clearly articulated in the UNECE Water Convention. In fact, the general commitments of member states relate mostly to environmental aims, from the prevention, control and reduction of water pollution to the need to adopt ecologically sound and rational water management.[139] Similarly, the Water Act 1974 is entirely devoted to addressing environmental aspects of water regulation. Interestingly, the Uruguayan Constitution asserts that water is an essential natural resource and a human right under a provision which concerns environmental protection.[140] In fact, the first principle guiding the development of water policy is the protection of the environment. The increasing importance of the environment in water law is also reflected in the ILA rules on water resources.[141]

The above exposition may give the impression that water law has effectively integrated environmental concerns. In reality, while some water law instruments are progressive, the majority is still far from effectively integrating an environmental perspective. At the international level, the UN Water Convention illustrates this point well. While the final negotiations took place in the aftermath of the United Nations Conference on Environment and Development that gave the environment significant visibility on the international stage, the Convention fails to clearly give environmental considerations priority over use.[142] At the domestic level, apart from the Water Act 1974, which is premised on environmental considerations, the rest of water law has remained more focused on considerations related to rules of access and control rather than conservation.

In the context of environmental law, a number of important developments have taken place. Some of the most basic principles of environmental law directly apply in the context of water. This is a logical extension of the fact that water has been a direct or indirect concern in environmental law from the outset.[143] While

[137] ch 3.B.1.

[138] Convention on the Law of the Non-navigational Uses of International Watercourses, New York, 21 May 1997, UN Doc. A/51/869, art 1.

[139] Convention on the Protection and Use of Transboundary Watercourses and International Lakes, Helsinki, 17 March 1992, UN Doc. ENWA/R.53, art 2.

[140] Constitución política de la República Oriental del Uruguay, art 47.

[141] Berlin Rules on Water Resources, International Law Association, Report of the Seventy-first Conference—Berlin (2004) c 5.

[142] eg P Wouters, The Legal Response to International Water Scarcity and Water Conflicts—The UN Watercourses Convention and Beyond (University of Dundee: Water Law and Policy Programme, 2003) 20.

[143] eg Declaration of the United Nations Conference on the Human Environment, Stockholm, 16 June 1972, UN Doc. A/CONF.48/14/Rev.1, principle 2.

nearly all environmental law principles are relevant in the context of water, some are of special relevance in the context of water law.

The principle of prevention has been at the centre of the development of environmental law, in fact characterizing the difference between a regulatory system that only addresses negative impacts after they have occurred and one that seeks to avoid significant environmental harm.[144] It seeks to ensure that measures are taken to avoid the known negative environmental impacts of planned activities. The basic nature of the principle of prevention means that its inclusion in water law has not proved too controversial.

The precautionary principle builds on the idea of prevention but adds an important new dimension. It suggests that measures must be taken to avert environmental harm even in situations where scientific knowledge is not conclusive as to the exact impacts of a planned activity.[145] It has been part of Indian law since the mid-1990s.[146] While no water legislation in India directly includes the precautionary principle, it is, for instance, one of the guiding principles of the UNECE Water Convention providing a basis to implement commitments to prevent, control, and reduce transboundary impacts.[147]

Equity is also of primary importance in environmental law. It includes two main dimensions. Intra-generational equity considers ways in which environmental regulation disproportionately impacts different groups. At the national level, it highlights the necessity to consider not only aggregate measures but also the ways in which specific individuals or groups benefit or suffer from environmental regulation.[148] At the international level, it considers the special needs of countries that have lesser resources to address environmental problems and countries that suffer more from environmental harm because of poverty. The concept of differential treatment, which recognizes the need for provisions which do not apply in the same way to all countries is the broader emanation of the principle of common but differentiated responsibilities recognized, for instance, in the Rio Declaration.[149] The second dimension, inter-generational equity brings an important new dimension to regulatory measures in providing a tool for taking into account the consequences of activities undertaken today in the long term.[150]

[144] eg P Sands, *Principles of International Environmental Law* (Cambridge: Cambridge University Press, 2003) 247–9.

[145] Rio Declaration on Environment and Development, Rio de Janeiro, 14 June 1992, UN Doc. A/CONF.151/26/Rev.l (Vol. l), principle 15.

[146] *Vellore Citizens' Welfare Forum v Union of India* (1996) 5 SCC 647 (Supreme Court of India, 1996).

[147] Convention on the Protection and Use of Transboundary Watercourses and International Lakes, Helsinki, 17 March 1992, UN Doc. ENWA/R.53, art 2(5)(a).

[148] On environmental justice, eg D Schlosberg, *Defining Environmental Justice—Theories, Movements, and Nature* (Oxford: Oxford University Press, 2007).

[149] Rio Declaration on Environment and Development, Rio de Janeiro, 14 June 1992, UN Doc. A/CONF: 151/26/Rev. 1 (Vol. l), principle 7.

[150] eg C Redgwell, *Intergenerational Trusts and Environmental Protection* (Manchester: Manchester University Press, 1999).

Finally, environmental concerns have also provided the basis for developments that span environmental law and other fields of law. This is, for instance, the case of the development of the human right to a clean environment. The right still faces resistance at the international level but has been widely adopted at the national level, including in India.[151] It links environmental standards to the impacts they have on each individual and casts the debate in terms of fundamental rights. The bridge between environmental law and human right which has been built over the past couple of decades through the right to a clean environment is significant in the case of water. Indeed, in a context where water and environment are increasingly linked and in a context where the right to water is the object of increasing attention, the right to a clean environment provides the third leg of a triangular relationship. The three legs, environment, water, and human rights, are all equally important and must be given similar importance.

This review indicates that the links between environmental law and water law have been made. Further, existing links have been integrated into water law instruments in some cases. As illustrated in the next chapter, environmental concerns have also become a central component of proposed reforms. Yet, a lot more remains to be done to effectively integrate environmental concerns into water law. Some possible leads are explored in chapter 6.

D. Towards Water Law Reforms

Water law has evolved and developed over time in response to changing conditions in the water sector. Thus the formal recognition of the human right to water has completely changed the conceptual framework that informs water law in general. Yet, water law is unadapted to the challenges of this century. This can be ascribed to a number of factors. Firstly, water law has developed in a sectoral manner and there is still no overarching framework water law. This is damaging because water law fails to reflect the unitary nature of water and the need to address all water challenges together. Secondly, despite a process of evolution some parts of water law such as irrigation law is by now conceptually too old to be able to effectively address today's challenges. Significant changes to existing irrigation acts are, for instance, necessary. Thirdly, existing rules governing access to water are in large part socially inequitable and environmentally unsustainable because of the link between land ownership and control over water. Fourthly, the existence of different legal principles governing different water bodies—such as surface and groundwater—despite the connexions between the two ensures that

[151] MR Anderson, 'Individual Rights to Environmental Protection in India', in AE Boyle & MR Anderson (eds), *Human Rights Approaches to Environmental Protection* (Oxford: Clarendon, 1996) 199.

there is significant potential for conflict between the different rules in place. Fifthly, while the scope of water law has significantly broadened over the past few decades, for instance, with the formal recognition of the human right to water, this remains largely inoperative because existing water laws fail to operationalize the right.

The above list indicates that there are a number of good reasons for reforming existing water law to ensure that it is adequate for today's challenges and operationalizes major conceptual changes brought about by the integration of human right or environmental concerns. There is relatively little dispute concerning the fact that reforms are necessary. There is, however, no consensus on the reasons why reforms are necessary and the types of reforms that should be introduced.

In practice, water law reforms that have been introduced since the 1990s are directly linked to water sector reforms. This specific basis for ongoing water law reforms has two main consequences in the context of this book. Firstly, since ongoing water law reforms are based on the principles of water sector reforms, the latter need to be introduced in order to understand the former. Chapter 3 thus analyses the underlying principles of water sector reforms. Water law reforms are then discussed in the next two chapters. Secondly, since the principles of water sector reforms have already been discussed among policy-makers, they are often not the object of much debate when they are integrated in legal instruments. The influence of law conditionality imposed by development banks is an additional reason explaining that alternatives to the model proposed under water sector reforms are not necessarily comprehensively considered. Chapter 6 consequently discusses some alternative bases for water law reforms. This confirms that there is more than one solution to the challenges identified, something which is given little visibility in ongoing policy and academic debates.

3

From Water Sector Reforms to Law and Policy Reforms

This book focuses on water law. Ongoing water law reforms are, however, not free-standing insofar as they are deeply influenced by the principles underlying what are known as water sector reforms. In other words, water sector reforms provide the policy context which informs ongoing water law reforms. An in-depth understanding of the principles underlying water sector reforms is thus necessary as an introduction to the analysis carried out in the next two chapters.

This chapter examines the notion of water sector reforms in its general and specific context. It first introduces the main characteristics of water sector reforms and the main underlying principles that inform the reforms. It then turns to the framework that guides reforms and introduces the type of law and policy changes that have been adopted. This chapter is by necessity partly general in scope. Indeed, water sector reforms are not India-specific. Uncertainty surrounds the origin of the reforms in India with some actors insisting on the domestic nature of the reforms. Regardless of who first pushed for the adoption of a new water agenda, there is striking convergence between the principles proposed at the national and international levels. Additionally, in recent years at least, water law conditionality in development aid projects is clearly identifiable and confirms at the very least the major role played by international institutions in the spread of water sector reforms. As a result, it is impossible to provide a comprehensive analysis of water sector reforms and water law and policy reforms in India without examining developments at the international level.

The exact extent of influence of international policy instruments and international institutions on water sector reforms is difficult to evaluate. However, there is little doubt that international institutions such as the World Bank and the Asian Development Bank (ADB) have had an important role in the adoption of water sector reforms and water law reforms in India. Their water policies are thus also considered here. This is, in fact, of topical relevance since water has been and will likely remain for the foreseeable future one of the important areas of interventions of development banks in India.[1]

A. Water Sector Reforms and Integrated Water Resources Management

The promotion of water sector reforms is informed by the idea that the water sector needs a major shake-up. Water sector reforms thus constitute a comprehensive programme of changes that are meant to impact all water uses. The concept of water sector reforms is extremely broad and applies in principle to any measures taken to effect changes in the water sector. Indeed, this is, for instance, the impression given by the definition of the ADB: 'Water sector reform refers to the whole of a country's policies, planning, implementation, and supporting activities to develop and manage its water resources and deliver water services to all users'.[2] Yet, in practice, what is known as water sector reforms refers to a much more specific set of measures and principles. Thus, reforms are premised on the need to address the current water crisis, which is seen as essentially a crisis of governance that needs to be addressed through reforms in water resources management.[3]

The focus on water resources management has been taken up in the context of the notion of integrated water resources management (IWRM). IWRM constitutes to a large extent the framework that informs water sector reforms. Policy-makers have identified it in recent years as 'the only way forward'.[4] One of its generally accepted definitions is that of 'a process which promotes the co-ordinated development and management of water, land and related resources, in order to maximize the resultant economic and social welfare in an equitable manner without compromising the sustainability of vital ecosystems'.[5] This is general enough to be acceptable by most people as a point of departure. However, the concept suffers from the same shortcomings that have been identified with the notion of sustainable development. It neither has clearly defined contours nor are there specific legal consequences that are attached to it. IWRM has thus been defined as an ideal concept rather than a set of specific guidelines and practices.[6]

[1] eg World Bank, Country Strategy for India (2004) 47–8.

[2] Asian Development Bank, What are Water Sector Reforms?, available at <http://www.adb.org/Water/CFWS/Water-Sector-Reforms/default.asp>.

[3] ibid.

[4] United Nations, *Water—A Shared Responsibility* (Paris: UNESCO, 2006) 12.

[5] Global Water Partnership—Technical Advisory Committee, Integrated Water Resources Management (Stockholm: Global Water Partnership, TAC Background Paper No. 4, 2000) 22.

[6] CJ Bauer, *Siren Song: Chilean Water Law as a Model for International Reform* (Washington: Resources for the Future, 2004) 9.

The proposal for IWRM has been linked to attempts to remedy identified shortcomings of existing water sector practices. In particular, IWRM advocates the need for a comprehensive view of water which avoids a sector-by-sector approach.[7] Thus, IWRM promotes basin-wide water planning, something that most people agree with in principle. Similarly, IWRM seeks to move beyond the consideration of water in isolation from environment and economic factors. It also promotes coordinated management and development of land and other resources.[8] This includes, for instance, links with the power sector since the increasing importance of groundwater makes electricity a prime determinant of access to water, leading to calls for linking reforms in the power and the water sectors.[9] Linking the two can be problematic if this is taken as implying that the same reforms introduced in the power sector should be adopted for water given their different nature. This is not a theoretical concern since the first water regulatory authorities introduced in India are partly inspired by the model of the earlier electricity reforms.[10]

Yet, IWRM is much more than a simple attempt to take a comprehensive view of water, it also seeks to comprehensively rethink its management. A number of management-related points have a central place in IWRM. In fact, notwithstanding the broad framework for IWRM, its central thrust is on issues related to the management of water. This includes various issues. Firstly, IWRM focuses on the development of participatory planning and implementation processes. This is meant to offer water users more of a say in decisions related to water resources. Secondly, IWRM calls for the decentralization of decision-making in reaction to the perceived failure of national administrative structures to deliver appropriate benefits to users. It thus involves the setting up of new institutions for the direct management of water demand, as illustrated by the case of water user associations.[11] Thirdly, it calls for a shift from supply to demand management. This highlights IWRM's emphasis on reducing the power and involvement of the state in the management of water resources in favour of users and private sector actors. In other words, IWRM is specifically premised as a vehicle for moving away from command and control regulatory approaches.[12]

IWRM focuses on the management of water resources within the context of a limited supply of water to ensure efficiency and equity without depleting the

[7] eg United Nations (n 4 above) 13.

[8] United Nations (n 4 above) 13.

[9] eg R Bhatia, 'Water and Energy Interactions' in J Briscoe & RPS Malik, *Handbook of Water Resources in India—Development, Management and Strategies* (New Delhi: The World Bank and Oxford University Press, 2007) 206.

[10] cf NK Dubash, 'Independent Regulatory Agencies: A Theoretical Review with Reference to Electricity and Water in India' (2008) 43/40 *Economic & Political Weekly* 43.

[11] T Shah, 'Institutional and Policy Reforms' in J Briscoe & RPS Malik, *Handbook of Water Resources in India—Development, Management and Strategies* (New Delhi: The World Bank and Oxford University Press) 306.

[12] eg United Nations Environment Programme, *Global Environment Outlook—GEO4—Environment for Development* (Nairobi: UNEP, 2007) 154.

resource.[13] Yet, despite this nominally broad, all-encompassing framework, IWRM does not provide a solution to all problems in the water sector. In fact, IWRM focuses on a specific number of issues and principles.[14] IWRM is, for instance, not comprehensive in its understanding of water because it largely reduces water to an economic good. This was, for instance, clearly laid out in an early version of chapter 18 of Agenda 21 stating that IWRM 'is based on water as a natural resource and an economic good'.[15] The final version is less specific and states that IWRM 'is based on the perception of water as an integral part of the ecosystem, a natural resource and a social and economic good'.[16] IWRM remains nevertheless deeply influenced by an economic logic as evidenced by the use of the term 'social good'. This is conceptually different from a human rights perspective on water.

Additionally, in practice IWRM's focus on management does not give other issues the same degree of importance. This is not unexpected given the focus on 'integrated management' but means that management is given disproportionate importance. Further, efficiency has been much more emphasized than equity.[17] More generally, the economic dimension of IWRM takes precedence. Thus, the pricing of water to ensure efficient water allocation and the creation of property rights have been given great importance.[18] Similarly, the call for managing water at the lowest appropriate level is not understood primarily as a measure for democratizing the water sector but as a way to achieve 'greater reliance on incentives, prices and markets'.[19]

IWRM has been embraced by the international community as the standard for moving forward in the water sector.[20] It has been used as the basis for interventions in the water sector in various parts of the world. However, its relevance for all countries has been questioned in its present form. Some proponents of IWRM have, for instance, criticized its actual implementation. The problems

[13] United Nations Development Programme, *Human Development Report 2006—Beyond Scarcity: Power, Poverty and the Global Water Crisis* (New York: UNDP, 2006) 153.

[14] eg NC Narayanan, 'Integrated Water Resource Management' (2008) 1 *Water Moves* 2, 4, arguing that the three intrinsic elements of IWRM are river basin management, stakeholder participation, and privatization/liberalization.

[15] Preparatory Committee for the United Nations Conference on Environment and Development, Options for Agenda 21—Protection of the Quality and Supply of Freshwater Resources: Application of Integrated Approaches to the Development, Management and Use of Water Resources, Third Session, Geneva, UN Doc. A/CONF.151/PC/42/Add.7 (1991) 13.

[16] Agenda 21, Report of the UNCED, Rio de Janeiro, 3–14 June 1992, UN Doc. A/CONF.151/26/Rev.1 (Vol. 1, Annex II) c 18(8).

[17] UNDP (n 13 above) 155.

[18] eg Shah (n 11 above) 306–7.

[19] International Conference on Water and the Environment, Report of the Conference (Geneva: World Meteorological Organization, 1992) 15.

[20] eg Plan of Implementation of the World Summit on Sustainable Development, Report of the WSSD, Resolution 2 (Annex), Doc. A/CONF.199/20 (2002) s 25. Also United Nations Environment Programme, *Global Environment Outlook—GEO4—Environment for Development* (Nairobi: UNEP, 2007) 119.

identified include the fact that, in practice, the broad aims of IWRM are reduced to a limited set of measures focusing on demand management.[21] These include the introduction of a water policy and a water legal framework, the recognition of river basins as the appropriate unit for managing water resources, the treatment of water as an economic good and the creation of tradable water rights, and the promotion of participatory water resource management.[22] Shah and Koppen argue that this set of prescriptions is failing in developing countries because the success of IWRM seems to be related to the growth and maturity of a country's economy.[23] In other words, countries like India where most people self-provide water and where access to water is still largely informally organized are not yet ready for the type of demand management strategies that IWRM requires. IWRM is thus deemed inappropriate at present, even by someone like Shah who seems to agree with the basic premises of the need to turn water into an economic good, to introduce property rights and to manage water at the basin level.[24] The current understanding of IWRM is also open to challenge because the broad framework it suggests is one that fails to make the links between water, land resources and social and economic development.[25] The issue is not with the notion of IWRM but the narrow economic approach that has informed its implementation. This is one of the lessons of Chile that was an early adopter of water sector reforms.[26]

B. Principles for Water Sector Reforms

Water sector reforms seek to be comprehensive. Indeed, if the reforms that are being proposed are fully implemented, this will lead to a complete transformation of the water sector. Yet, the focus on comprehensiveness hides the fact that ongoing water sector reforms are informed by a relatively narrow set of concerns and objectives. This section analyses some of the main principles underlying the introduction of water sector reforms. This provides the basis for understanding the exact nature of ongoing reforms and for distinguishing 'water sector reforms' from other reforms in the water sector that may be proposed.

Water sector reforms have two different meanings. The first relates to the set of prescriptions and principles that constitutes the current international policy-making consensus on the matter. The second refers to any reform in the water sector and thus has a much broader meaning. This section focuses on the first

[21] IWMI–TATA, IWRM Challenges in Developing Countries: Lessons from India and Elsewhere (IWMI, Water Policy Briefing, 2007).

[22] T Shah & B van Koppen, 'Is India Ripe for Integrated Water Resources Management? Fitting Water Policy to National Development Context' (2006) 41/30 *Economic & Political Weekly* 3413.

[23] ibid 3420.

[24] Shah (n 11 above) 316.

[25] V Upadhyay, Water-Forest Management, Law and Policy in Uttaranchal—Issues, Constraints, Opportunities (on file with the author, 2006).

[26] Bauer (n 6 above) 131.

narrower meaning because it informs a great number of reform efforts in many countries. It also constitutes the bedrock of most ongoing reforms in India. Other existing and possible reforms are highlighted in later chapters in the specific context of India. The fact that there are reforms which go beyond the scope of what is currently defined as water sector reforms implies that a positive or negative assessment of the latter does not imply that reforms are per se negative or positive.

This section introduces the principles underlying water sector reforms even though most of them have little legal content. This is due to the fact that it is, in fact, the same principles that constitute the bedrock of water sector reforms and water law reforms. This highlights some of the inherent limitations of water law reforms conceived mostly in relation to water sector reforms rather than by first analysing existing legal arrangements and determining the kind of changes that should be introduced.

1. Conservation

Chapter 2 highlighted the fact that environmental concerns are increasingly important in water law. In the context of water sector reforms, environmental issues are playing a much more central role. Indeed, there has been a significant shift in the past couple of decades from water policies linked mainly to health and development concerns to the idea that the need for water sector reforms is largely linked to environmental considerations.[27]

The central role given to environmental considerations is linked to the fact that water sector reforms are largely premised on ongoing and forthcoming water scarcity. Water scarcity is usually not defined as absolute scarcity of freshwater, which is not yet a global concern. Rather, what is targeted is the increasing scarcity of clean freshwater, whether due to excessive withdrawals or to decreasing quality. Scarcity and misuse of water are thus seen as threats to the protection of the environment.[28]

Environmental considerations in water acquired prominence in the early 1990s at the same time as a number of significant developments were taking place in international environment policy and law. The Dublin Statement highlights this shift by putting the essential quality of water as sustaining life on earth at the centre of the first principle for reforms.[29]

The primacy of the environment in water sector reforms is significant because it reverses the priority accorded to development concerns. The underlying idea is that it is environmental factors that constrain freshwater availability.

[27] cf M Finger & J Allouche, *Water Privatization—Trans-National Corporations and the Re-Regulation of the Water Industry* (London: Spon Press, 2002).

[28] eg Introduction to the Dublin Statement on Water and Sustainable Development, International Conference on Water and the Environment, Dublin, 31 January 1992.

[29] Dublin Statement on Water and Sustainable Development, International Conference on Water and the Environment, Dublin, 31 January 1992.

Conservation and protection of water resources thus become key factors. One of the results of the emphasis on environmental factors is that physical scarcity of water is given more importance than social scarcity. This, in turn, influences the policy responses which are given in terms of operative principles for water sector reforms.

The need to conserve and protect water is the source of operative principles for water sector reforms rather than the outcome. This raises different questions. Firstly, the emphasis on conservation should in principle be a welcome addition to the set of principles on which water policy is based. In practice, however, the emphasis on environmental aspects partly displaces the prominence given earlier to socio-economic factors. This is of particular concern because the emphasis on conservation is not matched by a similar emphasis on the human rights dimension of water.

Secondly, the emphasis on environmental aspects is noticeable for the narrow framework within which it is conceived. Indeed, under water sector reforms, the environment serves mostly as a justification for the operative policy principles highlighted below that form the core of the reforms. The new emphasis on water conservation is not used as a plank for rethinking the environmental content of water policies. It is thus not surprising to find that water sector reforms do not incorporate important advances made in recent years in environmental law. For instance, recent environmental law principles such as the precautionary principle find no place in water sector reform policies. In fact, the World Bank argues that most practitioners 'believe that the application of the precautionary principle would be a recipe for paralysis'.[30] Similarly, the rapid development of a human right to a clean environment in many countries is not used to strengthen the human rights dimension of reforms in the water sector.

Overall, water sector reforms are based on the recognition that sustainable water management can contribute to the protection of ecosystems.[31] This is an important step towards giving environmental issues a central role in water policy and law. At the same time, water sector reforms do little more than use the environment as the starting point for a set of economic and policy measures that form the core of the reforms.[32] They thus sideline the more substantive contribution that the integration of the environment can bring to addressing water issues. This is due in part to the fact that sustainability is to a large extent equated with cost recovery and financial viability.[33]

[30] World Bank, Water Resources Sector Strategy—Strategic Directions for World Bank Engagement (2004) 46.

[31] eg Ministerial Declaration of The Hague on Water Security in the 21st Century, World Water Forum, 22 March 2000.

[32] eg Bonn Recommendations for Action, Conference Report, International Conference on Freshwater, Bonn, 3–7 December 2001, 27.

[33] D Hemson, 'Water for All: From Firm Promises to "New Realism"?' in D Hemson et al. (eds), *Poverty and Water—Explorations of the Reciprocal Relationship* (London: Zed Books, 2008) 13, 28.

2. Water as a basic need

Most policy documents that highlight the key principles for the implementation of water sector reforms highlight the importance of water as a basic need necessary for human survival. Thus, the Bonn Conference asserted the primacy of basic human needs over other water uses in the context of the need to allocate competing demands on an equitable and sustainable basis.[34]

The emphasis on water as a basic need provides an important distinction from a human right approach. Indeed, while basic needs can serve as a partial indicator of the realization of human rights, the two are distinct.[35] Basic needs can be conceived as a benefit or service that can be considered in isolation. In the case of human rights, an essential element is universal human dignity, something which is not included in a basic needs approach.[36] A similar distinction can be made in the context of water between a human right conception and an essential service perspective whose underlying logic is different.[37]

In general, water sector reforms documents tend to avoid human rights language.[38] In certain contexts, like in the Dublin Statement, the recognition of a 'basic right' is mentioned. However, this is not a human right since it is a basic right, which is encompassed within the broader recognition of water as an economic good.[39] Since the human right to water is by definition not subordinated to an economic perspective on water, the 'right' recognized under the Dublin Statement is of a different nature. The Bonn Recommendations for Action confirm that water sector reforms are not based and centred on the realization of the human right to water. The Recommendations rather assert that 'water is an economic and social good, and should be allocated first to satisfy basic human needs'.[40]

Additionally, under water sector reforms the fulfilment of basic human needs is subordinated to the economic dimension of water. This is symptomatic of the broader philosophy of the reforms and highlights the limitations of the 'comprehensive' approach that is meant to be their hallmark. Thus, instead of building on the human right to water and providing ways to implement the right, water sector reforms adopt a different language. The Bonn Conference Report suggests,

[34] Bonn Recommendations for Action, Conference Report, International Conference on Freshwater, Bonn, 3–7 December 2001, 26.

[35] L-E Pettiti & P Meyer-Bisch, 'Human Rights and Extreme Poverty', in J Symonides (ed), *Human Rights: New Dimensions and Challenges* (Aldershot: Ashgate, 1998) 157, 170.

[36] ibid 171.

[37] W Vandenhole & T Wielders, 'Water as a Human Right—Water as an Essential Service—Does it Matter?' (2008) 26/3 *Netherlands Q Human Rights* 391, 423.

[38] eg Ministerial Declaration of The Hague on Water Security in the 21st Century, World Water Forum, 22 March 2000.

[39] Dublin Statement on Water and Sustainable Development, International Conference on Water and the Environment, Dublin, 31 January 1992, principle 4.

[40] Bonn Recommendations for Action, Conference Report, International Conference on Freshwater, Bonn, 3–7 December 2001, 23.

for instance, that governments should focus on setting and enforcing 'stable and transparent rules that enable all water users to gain equitable access to, and make use of, water'.[41] This goes alongside the more general aim of water sector reforms to shift government responsibilities away from delivering towards providing a framework for delivery by other actors but does not provide answers concerning the realization of human rights.

3. Water as an economic good

This section started with environmental concerns that constitute one of the main premises for ongoing reforms and the focus on basic needs, which link water sector reforms with the broader poverty eradication agenda at the centre of all international development policy. The importance of the environment and basic needs in the development of water sector reforms notwithstanding, it is economic aspects that overwhelmingly dominate the water sector reform policy agenda.

The Dublin Statement sets out the principle that water is an economic good.[42] This simple statement has been the basis for most of the reforms that have been adopted over the past couple of decades. The need to give water an economic value is premised on the fact that absence thereof explains existing wasteful and environmentally damaging water uses. This targets, for instance, low irrigation water rates that do not reflect the cost of providing the water to farmers. The issue of water rates is not a new one and has been debated for many years. Thus, the 1972 Irrigation Commission suggested the revision of water rates.[43] However, its framework of reference was not simply the economic nature of water but economic considerations put within a broader social and food security context. Thus, it suggested, for instance, that rates should be related to benefits to the users and should not unduly burden users.[44]

The recognition of water as an economic good has different implications. Firstly, in view of the recognition that water is both economically valuable and finite, the focus is put on managing water demand and increasing the efficiency of all water uses.[45] This is not necessarily straightforward because, as identified by the World Bank, scarcity does not necessarily lead to conservation and economic efficiency.[46]

Secondly, the recognition of the economic value of water calls for the introduction of pricing of all water services. The only qualification is the consideration of

[41] ibid 25.

[42] Dublin Statement on Water and Sustainable Development, International Conference on Water and the Environment, Dublin, 31 January 1992, principle 4.

[43] Report of the Irrigation Commission (New Delhi: Ministry of Irrigation and Power, 1972).

[44] A Narayanamoorthy & RS Deshpande, *Where Water Seeps!—Towards a New Phase in India's Irrigation Reforms* (New Delhi: Academic Foundation, 2005) 98.

[45] Bonn Recommendations for Action, Conference Report, International Conference on Freshwater, Bonn, 3–7 December 2001, 26.

[46] World Bank, India—Water Resources Management Sector Review—Initiating and Sustaining Water Sector Reforms (Report No. 18356-IN, 1998) 24.

the needs of the poor and vulnerable.[47] Even this is not always taken for granted and the Dublin Conference only talked about making available 'affordable supplies' to the unserved poor.[48] Additionally, water is also turned into a good that can be traded.[49] This has broad-ranging implications since a tradable good cannot easily be at the same time the subject of a human right. The issue of the basic needs of the poor and vulnerable is thus conceived within the new definition of water as a good which has, on the whole, the same characteristics as other tradable goods. The qualification of water as a 'social and economic good' does not change this basic position but indicates the limitations of an approach that sees water as an economic good since the fulfilment of the domestic water needs of all human beings cannot be achieved with an exclusive focus on water as a good.

Thirdly, besides the pricing of water per se, water sector reforms also seek to introduce full cost recovery as an operating principle for all water sector activities.[50] This is conceived as a prescription that applies to all water uses and all water users. In the case of domestic water, the proposed justification is that existing practices often disadvantage the poor who actually pay more for their water than the rich because the rich benefit from partly subsidized municipal water supplies that are not offered to poor or non-regularized areas.[51] The application across the board of the principle of cost recovery raises serious concerns. Indeed, as indicated above, water sector reforms derive their justification from their apparent focus on the basic needs of the poor. In a situation where two-thirds of the people completely lacking access to water survive on less than $2 a day, it is economically—as well as ethically—impossible to expect that unserved populations can finance their own improved access to water.[52] Besides the case of people in absolute poverty, cost recovery also raises broader questions. In the case of irrigation water, for instance, raising water rates to the point where they cover operation and maintenance costs and/or infrastructure costs is acknowledged as being an economic impossibility. Indeed, where farmers are charged Rs 300 per hectare, the full cost recovery charge has been estimated at Rs 10,000 per hectare.[53] This would also have untold impact on farmers, their livelihood as well as important food sovereignty consequences. Additionally, it has been found that farm-level inefficiencies are

[47] Ministerial Declaration of The Hague on Water Security in the 21st Century, World Water Forum, 22 March 2000.

[48] International Conference on Water and the Environment, Report of the Conference (Geneva: World Meteorological Organization, 1992) 30.

[49] cf J Sohnle, *Le droit international des ressources en eau douce—Solidarité contre souveraineté* (Paris: La Documentation française, 2002).

[50] eg CJ Vörösmarty, 'Fresh Water' in R Hassan, R Scholes & N Ash (eds), *Ecosystems and Human Well-being: Current State and Trends—Millennium Ecosystem Assessment Series Volume 1* (Washington: Island Press, 2005) 193.

[51] cf United Nations, *Water for People—Water for Life* (Paris: UNESCO, 2003) 341.

[52] UNDP (n 13 above) 7.

[53] AD Mohile, 'Government Policies and Programmes' in J Briscoe & RPS Malik, *Handbook of Water Resources in India—Development, Management and Strategies* (New Delhi: The World Bank and Oxford University Press, 2007) 10, 28.

not the main problem and that water prices are not the most significant element driving irrigation demand.[54] A concept which is sensitive to the various factors at play and the various dimensions of water is thus necessary.

Fourthly, the recognition of water as an economic good has implications for other conceptions of water. Indeed, it is likely that existing conceptions of water will be affected by the recognition of water as an economic good to the extent that they cannot coexist. Thus, the conception of water as a public trust or a common property conflicts at least in part with the economic good conception.[55] A fortiori, there are difficulties in reconciling the human right to water with the conception of water as an economic good. This is, for instance, illustrated in the candid acknowledgement that in a context that seeks to achieve full cost recovery, basic sanitation becomes mainly a household concern because 'sophisticated piped sewage systems require an unrealistic level of investment to meet the needs of the poor'.[56]

The recognition of water as an economic good has a number of additional repercussions. Apart from fundamentally changing the conception of water and affecting existing water laws and policies in significant ways, the characterization of water as a good implies that it can be traded under the rules of the existing international trading system. Thus, from being a substance that could neither be owned nor traded, water as conceived under water sector reforms becomes a good like any other tradable commodity. In fact, the change has not been completely sudden since water under certain forms, such as bottled water, has been traded for many years. What is new is that water sector reforms make something that was an exception the norm. Thus, it is today often argued that water is a good under the definition given by the GATT even if it is not specifically mentioned.[57] In fact, the real question that trade lawyers grapple with today is not whether water is a tradable good but whether it also falls under the scope of trade in services. At present, it is not possible to generally assert that all water services fall under the scope of the GATS.[58] Further, member states have not focused on water services commitments.[59] Yet, nothing stops member states from making such commitments and it is likely that changes will occur in the future.[60]

[54] I Ray, '"Get the Price Right"—Water Prices and Irrigation Efficiency' (2005) 40/33 *Economic & Political Weekly* 3659.

[55] M Moench, 'Allocating the Common Heritage: Debates over Water Rights and Governance Structures in India' (1998) 33/26 *Economic & Political Weekly* A46.

[56] United Nations (n 51 above) 112.

[57] eg Sohnle (n 49 above) 80.

[58] cf A Hildering, *International Law, Sustainable Development and Water Management* (Delft: Eburon, 2004).

[59] eg M Paquin et al., Les accords sur l'investissement et les services et la gestion de l'eau dans les pays en développement—Défis et opportunités pour l'atteinte des Objectifs du Millénaire pour le développement en matière d'eau potable et d'assainissement (Cible 10) (Centre international Unisféra, 2004).

[60] A Lang, 'The GATS and Regulatory Autonomy: A Case Study of Social Regulation of the Water Industry' (2004) 7 *J Intl Economic L* 801.

4. Individual property rights

Chapter 2 highlighted some of the main forms of control over water. These include individual rights of access to surface and groundwater water. Most of these rights are linked to control over land. One of the implications of these land-based rights is that they cannot be delinked from the ownership of the land.[61] Water rights under this traditional system were thus not independently tradable. This has not stopped the development of mostly informal and local water markets whereby individual landholders draw more water than necessary for their own needs and sell the surplus. Such informal water markets have been in existence for a number of years in several states.[62]

In principle, existing water rights do not give a right of ownership over the water itself. Indeed, even in England and Wales where water services are completely privatized, private companies do not own water.[63] The dichotomy between rights concerned with water services and rights over the water itself which has traditionally been beyond appropriation survives in most places. In fact, official documents usually conceive the new water rights as usufructuary rather than as rights of ownership.[64]

Water sector reforms seek to build on the existence of individual entitlements to water in water law. They, however, suggest a new framework for water rights where rights to water are delinked from land ownership. This provides the basis for the establishment of tradable water rights, something that has existed in some countries for quite some time but is novel in this form in India, as well as in most countries of the South with the exception of a few countries like Chile.[65] In fact, these are referred to as 'modern' water rights and seen as constituting a 'radical re-ordering of the status quo'.[66] These rights are meant to foster more efficient allocation of water by ensuring that low-value water users voluntarily desist from their rights while at the same time providing a basis for those requiring additional water to acquire the rights from those using water for low value purposes.[67] The

[61] eg United Nations (n 4 above) noting at 61 that '[w]ater rights are inextricably linked to property'. Note however that under the doctrine of prior appropriation the linkage between land and water rights was already severed. eg S Hodgson, Modern Water Rights—Theory and Practice (Rome: FAO, FAO Legislative Study 92, 2006) 13.

[62] eg S Janakarajan & M Moench, 'Are Wells a Potential Threat to Farmers' Well-being? Case of Deteriorating Groundwater Irrigation in Tamil Nadu' (2006) 41/37 *Economic & Political Weekly* 3977 and N Chandra, 'The Evolving Institution of Groundwater Markets: A Model of Socially Embedded Exchange' (2004) 1 *Indian Juridical Rev* 111.

[63] KJ Bakker, *An Uncooperative Commodity—Privatizing Water in England and Wales* (Oxford: Oxford University Press, 2003).

[64] eg World Bank, Water Resources Sector Strategy—Strategic Directions for World Bank Engagement (2004).

[65] Bauer (n 6 above).

[66] Hodgson (n 61 above) 1.

[67] World Bank, Water Resources Sector Strategy—Strategic Directions for World Bank Engagement (2004).

establishment of large regulated water markets has been advocated for more than a decade by the World Bank for India as an essential tool to reallocate increasingly scarce water supplies to high-priority uses.[68] The introduction of such water rights is premised on the need to avoid the 'damaging economic consequences' of a system where water rights are not clearly defined.[69] The rationale is thus squarely linked to economic efficiency goals and social goals are generally absent.[70]

The introduction of tradable water rights has generated significant debates in most developing countries. This is linked to the fact that the experience gained, for instance, in the United States does not indicate that water markets are necessarily apt at fostering equitable outcomes. Indeed, the increase in aggregate welfare hides the fact that losers tend to be the poorer farmers.[71] In developing countries, the possibility that water markets will lead poor farmers to engage in distress sales of water rights in return for short term monetary gain is of particular concern.[72] In India, a framework for trading water entitlements has already been adopted in several states and more may soon follow.[73]

5. Decentralization and user participation

Decentralization and water user participation are cornerstone principles of water sector reforms. The need for user participation is premised on the failure of centralized schemes to deliver benefits to water users at the local level and the need to reform the governance of water infrastructure schemes. The focus has been on the participation of specific water users in the management of irrigation or drinking water schemes.

The call for some form of participation is not new and can, for instance, be traced back to the sixth plan that called for farmers in each command area to participate in the development and management of water resources.[74] Yet, despite a number of earlier initiatives in some states, it is only with the large-scale implementation of water sector reforms that this has been taken up more systematically.

Under water sector reforms, decentralization and participation are often referred to in one breath even though the two notions refer to different realities. In the context of water sector reforms, decentralization is often used to refer to the transfer of responsibilities to civil society and the private sector rather than to its usual meaning of the transfer of authority from a central to a local government. This can be traced back to relatively early documents such as the first World Bank water policy whose definition of decentralization included in the same sentence

[68] World Bank, India—Water Resources Management Sector Review—Groundwater Regulation and Management Report (Report No. 18324-IN, 1998) xvi.
[69] Hodgson (n 61 above) 24.
[70] The only exception to this trend seems to be South Africa.
[71] UNDP (n 13 above) 180.
[72] UNDP (n 13 above) 181.
[73] ch 4.B.2–3, pp 121, 124–5.
[74] Planning Commission, Sixth Five Year Plan (New Delhi: Government of India, 1980) c 10.

decentralization to local governments and transfer service delivery functions to the private sector and water user associations.[75] The much more recent 2006 UN water report still has a similar understanding of decentralization.[76] This explains that in water sector reform debates, participation and decentralization are often used interchangeably.

The interchangeability of the two notions is, in fact, one of the hallmarks of the reforms. Indeed, water sector reforms are premised partly on a distrust of the government. This logically extends to a distrust of the ability of the government at all levels to deliver. This is problematic for two broad reasons. Firstly, the dictionary meaning of decentralization refers to a process of democratic decentralization in the context of a constitutionally defined system of governance.[77] In India, an extensive framework of democratic governance exists from the local level to the national level. The unwillingness of water sector reforms to clearly emphasize this dimension finds an echo in the setting up of water user bodies at the local level that are sometimes dissociated from panchayati raj institutions. This leads to considerable difficulties because such water user bodies have no place in the existing system of governance and there is, for instance, no accountability framework that provides the necessary safeguards to their operation. Secondly, the lack of a clear reference to the need to anchor the process of decentralization in the existing system of democratic governance provides the basis for another departure from the more usual understanding of the term. Thus, under water sector reforms, economic efficiency is acknowledged as another rationale for decentralization.[78] As a result, the democratization of decision-making and economic efficiency are, to a large extent, put on the same level. This has the unfortunate impact of downgrading at the outset social, environmental, and other considerations to second-tier issues.

Besides general considerations concerning the definition of the terms used, their specific meaning in water sector reforms policy documents requires analysis. The introduction of participatory approaches in water development and management have been key principles of water sector reforms. Thus, the Dublin Statement calls for users, planners, and policy-makers to be involved. Involvement is understood as a process that leads to decisions being taken at the lowest appropriate level 'with full public consultation and involvement of users'.[79] The distinction between the public at large and water users is significant since, in many cases, water users represent only a small subset of the general public. Thus, under participatory irrigation management, it is usually only landed farmers who are deemed users of the irrigation scheme while under the new principles for

[75] World Bank, Water Resources Management—A World Bank Policy Paper (1993) 72.
[76] United Nations (n 4 above) 75.
[77] *OED* (revised edition), Catherine Soanes and Angus Stevenson (eds) (2005).
[78] United Nations (n 4 above) 75.
[79] Dublin Statement on Water and Sustainable Development, International Conference on Water and the Environment, Dublin, 31 January 1992, principle 2.

drinking water schemes in India, the users are the ones that pay part of the capital costs of the schemes.[80] In the latter case, even though each and every individual is a user of drinking water, the notion of participation introduced under water sector reforms does not necessarily include all of them, while participation under the panchayat system would.

Participation is also understood under water sector reforms as a way to make water users more responsible for water management.[81] Two points can be made in this regard. Firstly, while participation is in principle conceived broadly, the emphasis is really put on participation in water management. As indicated by the UN Commission on Sustainable Development, participation includes involvement of stakeholders in planning and management but does not compulsorily include participation in decision-making processes.[82] Secondly, participation includes the introduction of additional responsibility for the users. In other words, participation and decentralization are meant to bring certain benefits to water users but these come with a number of new responsibilities as well.

The understanding of participation and decentralization under water sector reforms can be compared with broader conceptions of these terms. Thus, participation has a meaning that extends much beyond its scope under water sector reforms. Participation is in fact a procedural human right, as recognized in a number of legal instruments. One of the most elaborate statements is found in the Aarhus Convention.[83] Participation is understood today as a broad-ranging right which extends from the planning of a project or activity to its realization and implementation as well as to the preparation of plans, policies, and legally binding instruments.[84] There is thus a direct relationship between participation and forms of participatory democracy as developed, for instance, in the context of the panchayati raj institutions. Participation also includes the right to have access to information, which constitutes a condition for effective participation and the right to have access to justice to ensure accountability in the participatory process.[85] International water law does not yet recognize a specific right to participation. Yet, the 2004 ILA rules on water resources acknowledge that participation is deeply ingrained in international law and article 18 of the rules provides for a general right to participate in water management decisions and a right to have access to relevant information.[86]

[80] ch 4.A and 5.B.2.

[81] United Nations (n 4 above) 75.

[82] Resolution 13/1, Policy Options and Practical Measures to Expedite Implementation in Water, Sanitation and Human Settlements, Commission on Sustainable Development, Report on the Thirteenth Session, 11–22 April 2005, UN Doc. E/2005/29-E/CN.17/2005/12, s A(a)(iv).

[83] Convention on Access to Information, Public Participation in Decision-Making and Access to Justice in Environmental Matters, Aarhus, 25 June 1998, UN Doc. ECE/CEP/43 [hereafter Aarhus Convention].

[84] In an environmental context, Aarhus Convention (n 83 above) arts 6–8.

[85] These constitute the other two 'pillars' under the Aarhus Convention (n 83 above).

[86] Berlin Rules on Water Resources, International Law Association, Report of the Seventy-first Conference—Berlin (2004) art 18.

On the whole, decentralization and user participation are narrowly understood under water sector reforms, as well as under international water law. This conditions a number of the ongoing water law reforms which use the same conceptual framework. As a result, broader dimensions of participation, in particular its human right basis, are sidelined.

6. Institutional reforms and privatization

Institutional reforms constitute a new approach to tackle perceived problems of the water sector.[87] The basic assumption is that existing forms of regulation in the water sector have failed to deliver expected benefits and that a complete paradigm shift is required. The first element of the reforms is to break away from a model where water was 'developed' rather than 'managed'.[88] Water sector reforms thus seek to move away from engineering responses, such as dam-building, in favour of a focus on the management and allocation of available water.[89] Further, proposed institutional reforms are meant to ensure that water sector reforms are entrenched and have lasting impacts.

Water sector reforms first emphasize the overbearing presence of the state in the water sector. They attribute the failure to deliver appropriate benefits in terms of water conservation and the fulfilment of basic water needs to the existence of large water bureaucracies that are unable to effectively respond to local needs. This constitutes the basis for proposals to fundamentally reorganize the roles of the main actors in the water sector. The premise is that the interests of all stakeholders should be included in the management of water resources.[90] This leads, for instance, to the suggestion that the government should move away from service provision and restrict itself to facilitating the process. The state is thus encouraged to divest itself from some of the functions it has traditionally performed. One of the salient initiatives has been the setting up of regulatory bodies that are meant to be insulated from politics by being institutionally detached from existing ministries or departments dealing with water.[91]

The proposed withdrawal of the state from some of the functions it is currently performing leaves a void that needs to be filled. Water sector reforms make different suggestions in this regard. As indicated in the previous section, one of the proposals is to foster the participation of selected water users in the

[87] eg Bonn Keys, Conference Report, International Conference on Freshwater, Bonn, 3–7 December 2001, 22 and RM Saleth, 'Understanding Water Institutions: Structure, Environment and Change Process' in S Perret, S Farolfi & R Hassan (eds), *Water Governance for Sustainable Development* (London: Earthscan, 2006) 3.

[88] World Bank (n 46 above) 6.

[89] S Hodgson, Land and Water—The Rights Interface (Rome: FAO, FAO Legislative Study 84, 2004).

[90] Ministerial Declaration of The Hague on Water Security in the 21st Century, World Water Forum, 22 March 2000.

[91] ch 4.B.

management of water infrastructure. Another suggestion is to fill the gap left by the state with forms of private sector participation.[92] Proposed changes take several forms. They range from the general introduction of cost recovery to all water activities to the full privatization of water services. All of these are premised on their capacity to foster more economically efficient uses of water. Reforms are also often premised on the belief that only the private sector industry can provide solutions to existing water problems.[93]

In a narrow sense, privatization refers to the transfer of ownership through divestiture of state assets.[94] This is not a common model but is the reform model that was adopted in England and Wales at the end of the 1980s. The privatization in England and Wales led to the termination of a system based on the goal of universal provision with pricing based on social equity among households as well as between regions.[95] Instead, the public monopoly was transferred to corporations providing water supply on the basis of a private monopoly.[96]

A number of other models that do not involve the transfer of state assets have been used in other parts of the world. These include short-term service contracts where a private operator takes responsibility for a specific task such as reading meters or repairing leaks; management contracts where some operation and maintenance responsibilities are transferred; and much more extensive contracts such as build-own-operate-transfer contracts and concession contracts where the private operator manages the entire utility for a longer period and is required to invest in the maintenance and expansion of the system at its own risk.[97] These different forms of private sector participation also constitute forms of privatization as decision making responsibility and power over the assets is transferred to a private company.

Other reforms that cannot be qualified as privatization in the usual sense are also relevant here. These include the corporatization and commercialization of operations run by public utilities, which are often referred to as a form of creeping privatization.[98] This usually involves dissociating water service provision from

[92] eg World Bank, Uttar Pradesh Rural Water Supply and Environmental Sanitation Project (Ln 4056-IN) Terms of Reference for Rural Water and Environmental Sanitation Sector Study (2001).

[93] N Prasad, 'Privatisation Results: Private Sector—Participation in Water Services After 15 Years' (2006) 24/6 *Development Policy Rev* 669.

[94] DA McDonald & G Ruiters, 'Introduction: From Public to Private (to Public Again?)' in DA McDonald & G Ruiters (eds), *The Age of Commodity—Water Privatization in Southern Africa* (London: Earthscan, 2005) 1.

[95] Bakker (n 63 above) 5.

[96] On the implications of this kind of privatization, p 126.

[97] N Johnstone & L Wood, 'Introduction' in N Johnstone & L Wood (eds), *Private Firms and Public Water—Realising Social and Environmental Objectives in Developing Countries* (Cheltenham: Edward Elgar, 2001) 1, 10–11 and J Budds & G McGranahan, 'Are the Debates on Water Privatization Missing the Point? Experiences from Africa, Asia and Latin America' (2003) 15/2 *Environment & Urbanization* 87, 90.

[98] DA McDonald & G Ruiters, 'Theorizing Water Privatization in Southern Africa' in DA McDonald & G Ruiters (eds), *The Age of Commodity—Water Privatization in Southern Africa* (London: Earthscan, 2005) 13.

other services administered by the same utility and imposing market principles on the operation of water services. This has the effect of removing any cross-subsidies that may ensure the provision of free or cheap water and ensures that the public utility must function on the basis of cost recovery principles. These reforms have the impact of commodifying water which in turn makes actual privatization of water services easier to introduce subsequently.

Private sector participation is usually encountered in the context of the provision of water services, mostly in urban areas. Other forms of privatization have nevertheless also developed in recent years. This includes cases where the term privatization may not be the most appropriate term strictly speaking but where its use helps in understanding general policy trends. This is, for instance, the case of one of the few known cases of privatization of a river in the state of Chhattisgarh in India. In this case, a rather straightforward private sector participation in providing water to industrial units became the centre of a major controversy.[99] This was due to the fact that the company not only built a dam to provide water to users but also asserted rights over fishing in the area close to its dam and asserted the right to stop farmers living near the river from pumping water from the river.[100] This form of appropriation of flowing surface rivers which are without any doubt beyond appropriation by any individual in India has been strongly challenged. The Public Accounts Committee of the Chhattisgarh Assembly even recommended it should be cancelled.[101] This has, however, not happened yet.[102]

Private sector participation has been controversial for a number of reasons. Firstly, there is significant resistance to the broader process of commodification of water. Secondly, water does not lend itself easily to privatization since in practice, it is not feasible to expect more than a single provider to provide water in a given area. This quality of a 'natural monopoly' implies that there is no actual competition in the delivery of water. Competition takes place at the level of the bidding for the contract and the conditions imposed at the time of bid can only be enforced by a strong regulator, not through market mechanisms.[103]

Thirdly, the desire to devolve certain functions to the private sector to foster better access to water is contradicted by the need to strengthen the regulatory functions of the public sector to ensure that conditions imposed on the private entity are effectively respected. This has proved to be partially manageable in the case of a country like the UK where the government is able to effectively regulate

[99] cf AK Singh, *Privatization of Rivers in India* (Mumbai: Vikas Adhyayan Kendra, 2004).

[100] B Das & G Pangare, 'Privatisation: In Chhattisgarh, a River Becomes Private Property' (2006) 41/7 *Economic & Political Weekly* 611.

[101] SANDRP, 'Public Accounts Committee of Chhattisgarh Assembly Call for Immediate Cancellation of Sheonath Privatize (sic) Water Supply Project and Initiation of Criminal Proceedings against Responsible Officials' (2007) 5/3 *Dams, Rivers & People* 5.

[102] AP Putul, 'Privatisation Unlimited: Rivers for Sale in Chhattisgarh' (2008) *Infochange News and Features*, available at <http://infochangeindia.org/200802186943/Water-Resources/Analysis/Privatisation-unlimited-Rivers-for-sale-in-Chhattisgarh.html>.

[103] UNDP (n 13 above) 10.

private entities. In developing countries, privatization has been fraught with difficulties, in particular where governments deal with multinational companies.[104]

Fourthly, experience with private sector participation in water services delivery until now show at best mixed results.[105] In terms of the delivery of additional infrastructure, estimates indicate the private sector has only been directly responsible for an additional 250,000 connections over the past 15 years in developing countries.[106] Additionally, in practice, new connections seem to be provided preferentially to richer users rather than to poorer users who need them most.[107] Further, private sector participation is often linked to significant water price increases and the UNDP has noted that it is indeed public utilities that provide the cheapest water.[108] This is problematic because water consumption varies little with incomes and price rises thus harm the poor disproportionately more than the rich.[109] The figures given are very high indeed since even in a country like the UK, the poorest 1 per cent of households pays over 10 per cent of their income on water.[110] Put differently, in countries with a high level of poverty public finance remains a requirement to effectively extend access to unserved populations whether the provider is public or private.[111]

Fifthly, commercialization and privatization have been criticized for their incapacity to balance efficiency and equity. The allocation of water to higher value-added uses is promoted as a way to deploy water where it generates higher returns. This would, for instance, call for transfers from agriculture to industry in certain situations. Yet, while this may be appropriate in certain situations in developed countries, in many developing countries this is an insufficient answer to a much broader problem. Indeed, studies repeatedly show lower levels of poverty in areas where irrigation has been introduced.[112] Water transfers therefore have broader consequences that a simple economic efficiency perspective does not integrate. Thus, the withdrawal of irrigation water in favour of industrial use is likely to affect the poor disproportionately more and thus contribute to increasing inequalities.

C. Policy Framework Guiding Water Sector Reforms

Water sector reforms are peculiar when looked at from a legal perspective. Indeed, at one level, there has been remarkably little 'law' in water sector reforms until the

[104] A recently adjudicated dispute opposed Tanzania and the company Biwater. See *Biwater Gauff (Tanzania) Ltd v United Republic of Tanzania*, 24 July 2008, ICSID Case No. ARB/05/22.
[105] UNDP (n 13 above) 10.
[106] Prasad (n 93 above) 682.
[107] eg UNDP (n 13 above) 93 concerning Buenos Aires and Jakarta.
[108] UNDP (n 13 above) 83.
[109] Prasad (n 93 above) 675.
[110] Prasad (n 93 above) 676.
[111] UNDP (n 13 above) 10.
[112] UNDP (n 13 above) 153.

use of water law reforms became a tool to foster water sector reforms in recent years. Yet, the policy documents adopted at the international and national levels to set the framework for the introduction of water sector reforms have been exceptionally well followed. In fact, if these documents were binding, they would rate as very successful in terms of compliance.

The first characteristic of the framework for water sector reforms is that it is largely built around a set of non-binding documents such as non-binding resolutions adopted at international conferences and water policies at the national level. The second characteristic is that the same principles found in these documents have also made their way into the policies of the two main development aid institutions active in India, the World Bank and the ADB. Since both institutions have made a number of water-related loans in the past couple of decades to India, this explains in part the extent to which water sector reform principles have permeated the policy framework at the national level. Additionally, this also explains in part why different Indian states have adopted remarkably similar water policies as most have borrowed funds from one or the other development bank.

1. Context and nature of international policy documents

A number of international policy documents steer ongoing water sector reforms, a reflection of their importance throughout the world. Yet, there is no specific binding instrument that guides the reforms and fosters their implementation. At first sight, this is surprising because water sector reforms are momentous and there is an expectation that their main principles would be the subject of a negotiated instrument that would clearly lay out what all states agree on.

In reality, there have been no negotiations for an international agreement on water sector reforms. This can be attributed to several factors. Firstly, as noted in chapter 1, international water law is comparatively underdeveloped and states took many years to agree on what is essentially a framework convention whose scope is limited to shared watercourses. Additionally, the 1997 Convention has not entered into force. Secondly, water sector reforms are largely concerned with the management of water resources at the national level. This tends to make the intervention of international law less obvious. This is not, however, a definitive statement on the matter since biological resources are just as much under the sovereignty of individual states and this did not preclude agreement on cooperation in their conservation and sustainable use under the aegis of the Biodiversity Convention. Thirdly, it is unlikely that the principles put forward in water sector reforms would meet with the approval of a majority of states if they had to be incorporated in domestic law as part of the implementation of an international treaty.

The absence of binding international law instruments concerning water sector reforms is even more striking when the corpus of non-binding instruments is examined more closely. With regard to substantive principles, it is possible to identify a number of documents that propose the same vision for water sector reforms, starting with the Dublin Statement. In fact, the basic principles identified in the previous section of this chapter are repeated with a high degree of similarity in the main documents that serve as reference points for water sector reforms. This is true of the series of declarations adopted at the end of the sessions of the World Water Forum, as well as documents arising out of the Bonn International Conference on Freshwater or the Plan of Action of the World Summit on Sustainable Development.[113] This is also the case of documents that are not water-focused such as the action plan adopted by the G8 in 2003.[114]

The same principles have been repeated in a number of documents arising from international meetings since the beginning of the 1990s. This seems to indicate an important degree of consensus on the direction that reforms in the water sector should take.[115] This apparent broad consensus nevertheless warrants further remarks, in particular concerning the seminal Dublin Statement adopted at the International Conference on Water and the Environment. The Dublin Conference was organized in the context of the preparations for the United Nations Conference on Environment and Development (UNCED) but was separate from the meetings of the Preparatory Committee for UNCED. This was due in part to the fact that UNEP and the WMO had planned on organizing a conference before international policy attention focused on the preparations for UNCED.[116] This led to a hybrid formula. On the one hand, the proposed conference was to act as the formal entry for issues related to water for UNCED.[117] On the other hand, representation in the conference was not organized according to the practice that the UN General Assembly followed, for instance, in the Preparatory Committee for UNCED.[118] Indeed, the conference was not attended by government representatives but by a diverse mix of people, focusing on expert participants.[119]

The choice of experts to attend the Dublin Conference was not inappropriate considering that it was meant to be a technical conference in the first instance.

[113] eg Fourth World Water Forum Ministerial Declaration, Mexico City, 22 March 2006, Conference Report of the International Conference on Freshwater, Bonn, 3–7 December 2001 and Plan of Implementation of the World Summit on Sustainable Development (WSSD), 4 September 2002, Report of the WSSD, Resolution 2 (Annex), UN Doc. A/CONF.199/20 (2002).

[114] Water—A G8 Action Plan, Evian, 3 June 2003.

[115] eg Summary Report of the Stockholm Meeting on the Founding of a Global Water Partnership (December 1995) 1.

[116] Letter from GOP Obasi to J Pérez de Cuéllar, No 37.760/H/S-118, dated Geneva, 23 October 1990.

[117] Preparatory Committee for the UNCED, Protection of the Quality and Supply of Freshwater Resources: Application of Integrated Approaches to the Development, Management and Use of Water Resources, UN Doc. A/CONF.151/PC/73 (1991) 3.

[118] eg United Nations General Assembly Resolution 44/228, United Nations Conference on Environment and Development, 22 December 1989, UN Doc. A/RES/44/228, II.1.

[119] Preparatory Committee for the UNCED, Protection of the Quality and Supply of Freshwater Resources: Application of Integrated Approaches to the Development, Management and Use of Water Resources, UN Doc. A/CONF.151/PC/73 (1991) 5.

What is more surprising is that a technical meeting attended mostly by experts adopted a policy statement that has come to be regarded as the definitive international water policy statement. The fact that the Dublin Statement had little legitimacy in itself was recognized from the outset. Indeed, the statement was only 'commended' to government representatives attending UNCED.[120]

The drafting history of chapter 18 of Agenda 21 reveals that several of the issues emphasized in the Dublin Statement were already present in earlier drafts of chapter 18. Thus, the draft available to the fourth and last session of the Preparatory Committee for UNCED prepared before the Dublin Conference did include the idea that water must be considered as an economic good. However, the formulation used clearly put the environment and human needs ahead of the economic dimension of water. It stated that

> [p]riority must be given to the sustenance of land/water ecosystems, with particular attention to wetlands and biodiversity, and the satisfaction of basic human needs for drinking water, health protection and food security. For any water utilization beyond this, freshwater resources have to be considered as an economic good with an opportunity cost in alternative uses.[121]

This philosophy still informs the language of chapter 18 of Agenda 21, which states that '[w]ater should be regarded as a finite resource having an economic value with significant social and economic implications reflecting the importance of meeting basic needs'.[122] This does not coincide with the language of the Dublin Statement that simply called for water to be considered as an economic good in all its dimensions. It is thus surprising that the principles contained in the Dublin Statement are today often referred to as the Dublin-Rio principles.[123] This would be of little consequence if these principles had been subsequently widely debated in UN forums. In practice, however, international water policy has evolved since 1992 largely through meetings organized outside of a UN context. This is problematic because, despite its shortcomings, the UN, its subsidiary bodies and specialized agencies are the only institutions that have the legitimacy to address the kind of issues discussed in the Dublin Statement. It also matters because the UN gives comparatively more visibility to smaller states, in particular least developed countries. These happen to be the countries that have the biggest stake in debates concerning water as a source of life and livelihood. The sidelining of the

[120] Introduction to the Dublin Statement on Water and Sustainable Development, International Conference on Water and the Environment, Dublin, 31 January 1992.

[121] Preparatory Committee for the UNCED, Protection of the Quality and Supply of Freshwater Resources: Application of Integrated Approaches to the Development, Management and Use of Water Resources—Options for Agenda 21 as Revised During the Informal Consultations at the Third Session of the Preparatory Committee, UN Doc. A/CONF.151/PC/WG.II/L.17/Rev.1 (1992) 3.

[122] Agenda 21, Report of the UNCED, Rio de Janeiro, 3–14 June 1992, UN Doc. A/CONF.151/26/Rev.1 (Vol. 1, Annex II) c 18(68).

[123] eg R Hoare et al, External Review of Global Water Partnership—Final Report (2003) 4.

UN must also be seen in the context of the broader trend in water sector reforms that seeks to reduce the influence of governments in the water sector.

Overall, the consensus around water sector reform principles has been built in large part outside the UN. The importance given to the Dublin Statement over the past 15 years is linked to the fact that it was forwarded to member states participating in UNCED. Yet, in the same way that UNEP's Principles on Shared Natural Resources were never actually endorsed by the General Assembly,[124] the Dublin Statement principles were not endorsed. In fact, the Rio Declaration does not include any mention of water. The absence of any direct endorsement is significant because it indicates a lack of consensus around these principles at the international level. It is made more significant by the fact that the legitimacy of the Dublin principles has been linked to what is perceived as a UN endorsement.

Additionally, UN bodies have not shown a propensity to adopt language that mirrors the Dublin Statement. The section on water of the Programme for the Further Implementation of Agenda 21 begins by indicating that water is 'essential for satisfying basic human needs, health and food production, and the preservation of ecosystems, as well as for economic and social development in general'.[125] This constitutes a very different starting point than the Dublin Statement. The Commission on Sustainable Development later adopted the same language and confirmed that 'the priority to be accorded to the social dimension of freshwater management is of fundamental importance'.[126]

The discrepancy between the language of water sector reforms and that of UN bodies can be partly explained by the fact the international policy agenda for water sector reforms has been closely linked to the involvement of several organizations that have little to do with the UN. In fact, two of the most influential organizations over the past decade were set up in 1996 with the specific agenda of promoting and fostering water sector reforms.[127] The World Water Council is usually described as a think-tank and is constituted in the form of an association under French law.[128] Its objectives include the development of 'a common strategic vision on integrated water resources management on a sustainable basis' as well as the promotion of 'the implementation of effective policies and strategies worldwide'.[129] One of its main activities has been the organization of the world

[124] United Nations General Assembly Resolution 34/186, Co-operation in the Field of the Environment Concerning Natural Resources Shared by two or More States, 18 December 1979, Resolutions and Decisions Adopted by the General Assembly During its 34th Session, UN Doc. A/34/46.

[125] United Nations General Assembly Resolution S/19-2, Programme for the Further Implementation of Agenda 21, 19 September 1997, UN Doc. A/RES/S-19/2, para 34.

[126] Commission on Sustainable Development, Decision 6/1, Strategic Approaches to Freshwater Management, Report on the Sixth Session, UN Doc. E/1998/29-E/CN.17/1998/20 (1998) para 3.

[127] The setting up of a world water council was already suggested at the Dublin Conference. International Conference on Water and the Environment, Report of the Conference (Geneva: World Meteorological Organization, 1992) 42.

[128] World Water Council Constitution, 14 June 1996 (as amended).

[129] ibid art 2(3).

water forums. The second is the Global Water Partnership (GWP) which was set up by the World Bank, UNDP and the Swedish International Development Agency.[130] The arrangement was formalized in 2002 with the setting up of a GWP Organization whose mandate is to support the GWP Network.[131] The GWP is based on the 'simple concept' that 'freshwater resources are finite and their various uses are interdependent, but most of the water management activities carried out at the national or international level do not recognize these interdependencies'.[132] This is reflected in statutes of the GPW Network which determine that the single objective of the Network is to develop and promote the principles of integrated water resource management.[133]

One of the objectives behind the setting up of these two new bodies has been to provide new platforms where a greater number of entities involved in the water sector can be involved, in particular private sector water companies.[134] One of the impacts has been to marginalize the role of the UN system in water policy through the emphasis of these new bodies that are deemed more effective because they are not limited to public sector actors.

The importance of the World Water Council and the Global Water Partnership has been reinforced by their cooperation. They have thus jointly contributed to an initiative focusing on the financial aspects of water sector reforms. This is in line with the decision of the World Water Council to make financing one of its top priorities. The first initiative to emerge was the World Panel on Financing Water Infrastructure chaired by Michel Camdessus. This panel published a report which has been key in further entrenching the idea that water is an economic good and that cost recovery is central to successful reforms, in particular to attract private sector financing.[135] The second report chaired by Angel Gurria focused on water financing and local governments and financing water for agriculture.[136] It built on the idea of proximity by arguing that water services are a local issue best provided by local entities and proposes, for instance, a series of measures to ensure finance flows directly to the local level.

The various documents, reports and resolutions of the past 15 years provide the impression that there is consensus among all actors involved in water around the proposed water sector reforms. A lot has been done to bolster the legitimacy

[130] eg R Hoare et al, External Review of Global Water Partnership—Final Report (2003).

[131] Statutes for the Global Water Partnership Network and the Global Water Partnership Organisation, 12 December 2002.

[132] S Özgediz & B Axelsson, Report of the Management Advisory Review of the Global Water Partnership (Stockholm: Global Water Partnership, 1998) 2.

[133] Statutes for the Global Water Partnership Network and the Global Water Partnership Organisation, 12 December 2002, art 2.

[134] eg R Petrella, *The Water Manifesto: Arguments for a World Water Contract* (London: Zed, 2001) 23 and Finger & Allouche (n 27 above) 28.

[135] World Panel on Financing Water Infrastructure, Financing Water for All (Marseilles: World Water Council, 2003).

[136] Task Force on Financing Water for All, Enhancing Access to Finance for Local Governments—Financing Water for Agriculture (Marseilles: World Water Council, 2006).

of the reforms through a careful use of words. Thus, the 1992 Dublin meeting was an 'international conference', the World Water Council organizes 'world forums' and sets up a 'world panel' on financial issues. Further, the lines between a global UN meeting and some of the much more narrowly conceived meetings on water sector reforms has been increasingly blurred since a selection of elected officials participate in the latter meetings.

The fact that the UN has not been given a strong leadership role to lead the community of states provides much more scope than at the national level for alternative or additional initiatives and institutions to emerge. Yet, the UN has a legitimacy which is unmatched by other existing institutions when it comes to addressing fundamental global human rights, social, and environmental problems. While nothing precludes the various initiatives that have been taken by different actors, these should not be indirectly given the legitimacy of a UN consensus. Indeed, water sector reforms are yet to be openly debated in the same level of detail within the UN system where the priorities of least developed countries and all other small developing countries are more likely to be heard. As a result, it is impossible to understand the existing package of documents as more than a series of proposals by a limited number of states and actors for sweeping reforms of the water sector. At present, these reforms do not have the sanction of a binding international framework. Similarly, there is no consistent body of UN General Assembly resolutions or similar soft law instruments that would provide specific indications of such a consensus among all UN member states.[137]

2. Role of international financial institutions and their policies

International financial institutions have been playing an increasingly important role in water sector reforms and water law reforms in developing countries. This is particularly true for the World Bank both in terms of financial commitments and in terms of policy making in developing countries.[138] The water policy of the World Bank has dramatically evolved over the past two decades. It is in the early 1990s that a shift away from water resource development and towards water resource management was undertaken.[139] This was driven by concerns for the sustainability of water uses. The proposed solution is better management of water, itself only possible if water is conceived as an economic good. It is also at this point that the Bank moved towards encouraging the

[137] Whereas some resolutions, such as United Nations General Assembly Resolution 50/126, Water Supply and Sanitation, 23 February 1996, UN Doc. A/RES/50/126 use the language of water sector reforms, most resolutions concerned with water do not engage with this. For the latter, eg United Nations General Assembly Resolution 56/192, Status of Preparations for the International Year of Freshwater, 2003, 7 February 2002, UN Doc. A/RES/56/192 and United Nations General Assembly Resolution 57/252, Activities Undertaken in Preparation for the Year of Freshwater, 2003, 21 February 2003, UN Doc. A/RES/57/252.

[138] Finger & Allouche (n 27 above) 43.

[139] World Bank, Water Resources Management—A World Bank Policy Paper (1993).

provision of water through private ownership and operation backed by an effective regulatory framework.[140] This coincided with the adoption of the first formal water policy. Over the past fifteen years, the Bank has been putting 'an overwhelming emphasis on more efficient service delivery and management of water'.[141]

The 1993 Policy Paper and the 2004 Water Resources Sector Strategy have provided the broad framework within which Bank interventions take place. The 1993 Policy Paper proposed a new approach to manage water. The basic framework proposed was to treat water as an economic good, to decentralize management and delivery structures, to put greater reliance on pricing and to foster stakeholder participation.[142] It also suggested that the Bank would foster the development of a 'strong legal and regulatory framework', an early indication of the type of interventions that have been undertaken by the Bank since then.[143]

The 2004 Strategy has updated the policy message in view of the experience with water sector reforms over the previous decade.[144] Under the current strategy, water plays a central role for the Bank. Several key elements are highlighted. Firstly, effective water resource development and management comprises broad-based interventions including major infrastructure like dams and inter-basin transfers. Secondly, the Bank reiterates that it identifies a link between poverty and environmental degradation and thus sees improving catchment quality as a way to address poverty. Thirdly, institutional reforms seeking to improve the performance of water-related institutions are deemed to be beneficial to everyone, including the poor. Fourthly, poverty-targeted interventions are seen as contributing to the realization of the Millennium Development Goals.

Bank policy is further specified in the context of the operational policy on water management.[145] The operational policy sets a series of clear priorities for World Bank engagement. It first emphasizes the need for designing water resource investments, policies, and institutions. These must be based on principles that achieve cost recovery, water conservation, and better allocation of water resources. Additionally, the involvement of water users in planning and management is promoted. The Bank specifically focuses on the division of responsibilities between public and private entities and considers that a variety of organizations, from community organizations to private firms contribute to what it conceives as

[140] Finger & Allouche (n 27 above) 71.

[141] RPS Malik, 'World Bank Policies and Lending Assistance' in J Briscoe & RPS Malik, *Handbook of Water Resources in India—Development, Management and Strategies* (New Delhi: The World Bank and Oxford University Press, 2007) 69.

[142] World Bank, Water Resources Management—A World Bank Policy Paper (1993) 10.

[143] ibid 13.

[144] World Bank, Water Resources Sector Strategy—Strategic Directions for World Bank Engagement (2004).

[145] World Bank, Water Resources Management, Operational Policy 4.07 (2000).

decentralization. Indeed, it supports private sector involvement in the provision of water and sanitation services.[146]

In the context of this study, one of the important elements of the Bank's water policy is its position on water law reforms. The Operational Policy stresses as one of its priorities the establishment of legal frameworks 'to ensure that social concerns are met, environmental resources are protected, and monopoly pricing is prevented'.[147] The Bank's Water Strategy addresses water law reforms in the context of its emphasis on the need for new sources of finance for water resources infrastructure. In this context, the Strategy argues in favour of 'effective and predictable rules and institutions for balancing the interactions of investors, government, and users and other affected people'.[148] The Bank specifically points out that only the public sector can develop this stable environment. The importance given to regulation is significant because it arises in a context of reforms that seek to divest the state and the public sector generally of many of its existing functions. In other words, the public sector is called upon to provide a new regulatory framework that gives powers to other actors while requiring the state to provide the framework within which the private sector and other non-governmental actors interact. This is understandably not an agenda that any government would undertake on its own. As a result, the World Bank sees part of its role as providing investments and assistance in developing legal and institutional arrangements.[149] The Bank thus directly acknowledges that it includes in its role the promotion of the development of new water laws at the national level. While the Strategy does not directly promote law conditionality, it indirectly justifies and fosters its use in Bank loans as a way to ensure that borrowing countries implement the proposed reforms.

Overall, the priorities of the Bank coincide largely with water sector reform principles identified earlier in this chapter. This is not necessarily surprising given that the Bank has been an active participant in the forums that have contributed to the establishment of the policy framework for water sector reforms at the international level. It had, for instance, a leading role in the setting up of the Global Water Partnership. It not only co-founded it but also provided a majority of its funding at the outset.[150]

In India, the two main relevant institutions are the World Bank and the ADB. Their internal policies applicable to the loans they make are based on the same broad principles. These policies have been increasingly influential in

[146] World Bank, Water Resources Sector Strategy—Strategic Directions for World Bank Engagement (2004) 19.

[147] World Bank, Water Resources Management, Operational Policy 4.07 (2000) para 2(F).

[148] World Bank, Water Resources Sector Strategy—Strategic Directions for World Bank Engagement (2004) 45.

[149] ibid 45.

[150] S Rana and L Kelly, The Global Water Partnership Addressing Challenges of Globalization: An Independent Evaluation of the World Bank's Approach to Global Programs (Washington, DC: World Bank Operations Evaluation Department, 2004) 22.

India where a majority of states have been involved in either a World Bank or ADB loan in the water sector since the early 1990s.[151] This explains, for instance, why Saleth finds that institutional reforms have been 'more due to purposive reform programmes than to any natural process of institutional evolution'.[152]

Indian states that have accepted international funding have had to comply with the financial institution's own policies, thereby influencing state level policy and law making. Additionally, some water sector projects include policy and law conditionality which directly calls on states to introduce new measures or laws or amend existing instruments. This is, for instance, the case of the ADB's Chhattisgarh Irrigation Development Project that imposed on the state of Chhattisgarh to enact an amended participatory irrigation management act and went as far as specifying some of the provisions that needed to be included in the amended act.[153] Similarly, the World Bank's Madhya Pradesh Water Sector Restructuring Project imposes on the state to 'prepare and submit for consideration for adoption' an appropriate draft enabling legislation for the establishment of a State Water Tariff Regulatory Commission.[154] The conditions also include a list of some of the functions of this Commission which will review and monitor water sector costs and revenues and ensure the setting of bulk water user fees to enable water sector operations to be financially viable.[155] These are significant developments because this reflects a direct involvement of international institutions in law making at the national level.[156] It also constitutes one new step in the kind of interferences that go alongside with the credit agreements that borrowing states must sign.[157]

Besides the amendment or adoption of laws, conditionality also arises in the context of less broad-ranging but potentially very significant measures that can be adopted by the concerned governments. The Kerala Urban Sustainable Development Project imposes, for instance, on the state government the conversion of existing standposts either to individual metered house connections or to metered standposts.[158] An even more brutal condition is the one that was imposed on

[151] For an update, eg Manthan Adhyayan Kendra, Database of Projects and Programs Involving Privatisation and Commercialisation, available at <http://www.manthan-india.org/article23.html>.

[152] Saleth (n 87 above) 3.

[153] Asian Development Bank, Report and Recommendation of the President to the Board of Directors on a Proposed Loan and Technical Assistance Grant to India for the Chhattisgarh Irrigation Development Project, Doc. IND 37056 (2005) 17.

[154] World Bank, Madhya Pradesh Water Sector Restructuring Project (Project Appraisal Document, Report No. 28560-IN, 2004) 10.

[155] World Bank, Madhya Pradesh Water Sector Restructuring Project (Project Appraisal Document, Report No. 28560-IN, 2004) 3.

[156] cf V Upadhyay, Law under Globalization—Assessing 'Donor Supported' Law Making and Judicial Behaviour in India (New Delhi: Social Watch Coalition, 2008).

[157] cf S Randeria, 'Glocalization of Law: Environmental Justice, World Bank, NGOs and the Cunning State in India' (2003) 51/3-4 *Current Sociology* 305.

[158] Asian Development Bank, Kerala Urban Sustainable Development Project No 32300, Report and the Recommendation of the President to the Board of Directors (2005) 19.

the governments of Rajasthan and Karnataka to implement a water supply disconnection policy while increasing water tariffs.[159]

The law and policy conditionality imposed on borrowing states largely mirrors the principles of the banks' own policy guidelines, which reflect, in turn, the main principles of water sector reforms. The desire of the banks to influence law and policy development at the national level is directly acknowledged in the case of the World Bank. The first of the two main functions of its water policy is to 'encourage reforms in water management institutions, policies, and planning in borrowing countries'.[160] Extending legislation and regulatory mechanisms and institutions is thus a new priority for the Bank.[161] The focus on state level changes is noteworthy. Indeed, while it corresponds with the constitutional position in India, it is also in line with a shift in the 1990s that saw the World Bank moving away from a focus on the Union government in favour of engagement with individual states.[162]

The role of the World Bank as a promoter of law and policy reforms is an offshoot of the shift in water policy over the past 15 years. This goes hand-in-hand with the move away from a project and engineering focus to putting the emphasis on management aspects. In the case of India, this has led to the Bank becoming 'increasingly involved in Indian policy debates on institutional reforms both within and outside the water sector'.[163] The Bank has, in fact, proffered specific policy recommendations for India, including with regard to water law. In its assessment of the problems that India is facing, it bemoans the fact that policy makers have focused on the adoption of laws and policies which are often not implemented.[164] It also finds that the public sector is ill-equipped to develop an enabling legal and regulatory framework which it sees as one of the remaining useful functions of the public sector in the water sector.[165]

An interesting feature of the World Bank's engagement with the water sector in India over the past two decades is that a number of its main messages have been or are in the process of being implemented. The comprehensive water sector review undertaken at the end of the 1990s is illustrative of this trend. The review

[159] eg Asian Development Bank, Rajasthan Urban Infrastructure Development Project No 29120, Report and Recommendation of the President to the Board of Directors (1998) 29 and Asian Development Bank, Karnataka Urban Development and Coastal Environmental Project No 30303, Report and Recommendation of the President to the Board of Directors (1999) 32.

[160] GK Pitman, Bridging Troubled Waters—Assessing the Water Resources Strategy (Washington, DC: World Bank, 2002) 45.

[161] U Hoering & AK Schneider, King Customer? The World Bank's New Water Policy and its Implementation in India and Sri Lanka (Stuttgart: Brot für die Welt, 2004) 9.

[162] eg R D'Souza, *Interstate Disputes over Krishna Waters—Law, Science and Imperialism* (New Delhi: Orient Longman, 2006).

[163] R Bhatia, 'Water and Economic Growth' in J Briscoe & RPS Malik, *Handbook of Water Resources in India—Development, Management and Strategies* (New Delhi: The World Bank and Oxford University Press, 2007) 99, 126.

[164] J Briscoe & RPS Malik, *India's Water Economy—Bracing for a Turbulent Future* (New Delhi: The World Bank and Oxford University Press, 2006) xx.

[165] Briscoe & Malik (n 164 above) xxi.

suggested, for instance, that constitutional provisions and water legislation did not provide an appropriate framework to address water sector problems.[166] It alluded, for instance, to the need to conceive water regulation at the basin level, the need to introduce clearly defined and secure commercially transferable water rights,[167] and the need for participation of local people in managing water infrastructure.[168] It also suggested a shift towards demand-oriented approaches, financial viability of water service delivery and a shift in the role of the government from a provider and financier to one of facilitator.[169] A noteworthy aspect of the Bank's engagement with policy changes is that they proceed from a process of cross-fertilization between national and international policy-making initiatives. Thus, the Bank justifies some of its policy prescriptions on the basis of principles stated in the five year plan.[170] This includes, for instance, calls for private sector participation which the Bank traced in 1998 to the eighth plan.[171]

With regard to water law reforms, it is symptomatic that the 1998 report specifically laments the absence of a 'solid legislative and regulatory framework'.[172] It thus calls for review and revision of existing laws and the adoption of new legislation. It does not escape notice that there has been a rapid acceleration in water law reform since 1998 in many states. While it is impossible to point out on the basis of existing written documents the exact extent of World Bank involvement in those changes, anecdotal evidence seems to suggest that its influence is much more significant than the idea of cross-fertilization between the national and international levels implies. In fact, the Bank is not necessarily shy about its role. The country strategy acknowledges, for instance, that it 'has helped to pilot and begin to scale up a reform program' for water services in rural areas.[173]

The Bank's understanding of water is also of great importance in the context of the law reforms that are being implemented. It sees water 'not [as] a national issue but an intensely local one'.[174] This corresponds to the focus on decentralization and participation in water sector reforms and is thus not particularly surprising in this context. Yet, this statement is at best unhelpful. Water is at the same time a local, block, district, state, inter-state, national, and international level issue. In reality, not even the World Bank believes that water is not a national issue. It uses a different lens to look at different issues that are in fact all

[166] World Bank, India—Water Resources Management Sector Review—Initiating and Sustaining Water Sector Reforms (Report No. 18356-IN, 1998) 13.

[167] ibid 14.

[168] ibid ii.

[169] ibid vii.

[170] World Bank, India—Water Resources Management Sector Review—Rural Water Supply and Sanitation Report (Report No. 18323, 1998) 2.

[171] World Bank, India—Water Resources Management Sector Review—Urban Water Supply and Sanitation Report (Report No. 18321, 1998) 29.

[172] World Bank, India—Water Resources Management Sector Review—Initiating and Sustaining Water Sector Reforms (Report No. 18356-IN, 1998) 56.

[173] World Bank, Country Strategy for India (2004) 40.

[174] Briscoe & Malik (n 164 above) 15.

part of the same coin. Whereas it advocates a focus on decentralization and participation for some water management related tasks, it promotes at the same time involvement in big supply-side water infrastructure, such as dams and inter-basin transfers. This does not need to be a contradiction from the point of view of the Bank. Indeed, it may argue that large-scale schemes are necessary to ensure water availability at the local level. Yet, the disconnect between the two is artificial. In fact, there is no reason to argue that water users should have a say in the use of water at the local level if that water can only be made available through a much larger scheme. Similarly, the focus of water sector reforms on decentralization and participation masks the return to the promotion of large-scale water infrastructure. In this sense, identifying water as a 'local' issue is a misnomer because it fails to provide an overall view of the different strategies being put in place in the water sector.

3. National water policies

The development of water policies in India has been closely linked to the introduction of water sector reforms and more recently water law reforms. Indeed, before 1987, despite the importance of water in all aspects of life and despite the existence of a number of water-related laws and regulations and case law, there was no water policy giving the water sector a unified direction. The National Water Policy 1987 was thus a landmark inasmuch as it constituted the first attempt to provide a general framework within which the water sector should operate.

The development of water policies is an area where international agencies have played at least an important catalytic role. The 1998 World Bank review of the water sector specifically lamented the fact too few states had formulated water policies and called on them to do so within the context of the National Water Policy.[175] Additionally, the Bank also suggested at the time that some of the features of the National Water Policy should be 'improved' to reflect the new economic policy put in place during the 1990s.[176] A revision of the National Water Policy was in fact adopted in 2002 and most state water policies have been adopted over the past decade.

Most water policies seek to establish basic principles for all activities related to water and ensure consistency throughout the various sectors. The National Water Policy is understandably the most important of the existing policies. The 2002 revision of the policy provides insights on the changes that have taken place in water policy thinking since 1987. The current version of the policy puts, for instance, stronger emphasis on the principles underlying integrated water resource

[175] World Bank, India—Water Resources Management Sector Review—Initiating and Sustaining Water Sector Reforms (Report No. 18356-IN, 1998) 56.

[176] World Bank, India—Water Resources Management Sector Review—Inter-sectoral Water Allocation, Planning and Management (Report No. 18322, 1998) 21.

management.[177] The focus on encouraging private sector participation is also an addition of the revised policy.[178] While a number of differences can be identified not only between the 1987 and 2002 policies but also between the various state policies, there is a broad uniformity with regard to the basic principles which are put forward. As a result, a number of commonalities can be identified in the existing policies.

a) General features of water policies

A number of general trends can be identified. Firstly, water policies conceive water as a natural or economic resource which can be harnessed to foster the productive capacity of the economy, from irrigation water for agricultural production to water for hydropower. The National Water Policy laments the fact that an insufficient percentage of the water is currently harnessed for economic development and calls for 'non-conventional' methods of water utilization such as inter-basin water transfers and seawater desalination as large-scale, high technology solutions to improve overall water availability.[179] This message is further reinforced in the recent World Bank report stressing that India has not developed enough big water infrastructure.[180]

Secondly, policies introduce prioritization of water uses. The national policy provides, for instance, that water should be allocated in the following order: drinking water, irrigation, hydro-power, ecology, agro-industries and non-agricultural industries, navigation and other uses.[181] Several state policies follow a relatively similar ordering scheme.[182] The main exception is the Maharashtra policy, which puts industrial use before irrigation. This appears to be a direct response to the World Bank's critique of the fact that industrial use is often the last priority even though value-added is usually greater than for water used in irrigation.[183] Additionally, in Himachal Pradesh while the environment is specifically mentioned as one of the priorities, something that all state policies do not do, tourism and environment are put as a single priority, thus sidelining the most important environmental issues that arise in the context of water.[184] Overall, there is a clear emphasis on drinking water as the first priority in water allocation.[185]

[177] See AD Mohile, 'Government Policies and Programmes' in J Briscoe & RPS Malik, *Handbook of Water Resources in India—Development, Management and Strategies* 10 (New Delhi: The World Bank and Oxford University Press, 2007) 10, 11.

[178] National Water Policy 2002, s 13.

[179] National Water Policy 2002, s 3(1–2).

[180] Briscoe & Malik (n 164 above) 30.

[181] National Water Policy 2002, s 5.

[182] eg Rajasthan State Water Policy 1999, s 8, Uttar Pradesh Water Policy 1999, s 5(1) and Kerala State Water Policy 2007, s 2(1).

[183] World Bank, India—Water Resources Management Sector Review—Inter-sectoral Water Allocation, Planning and Management (Report No. 18322, 1998) 21.

[184] Himachal State Water Policy 2005, s 7.

[185] Kerala State Water Policy 2007, s 2(1) uses the broader term 'domestic use'.

This is reinforced in some policies by a call for the government to provide adequate safe drinking water facilities to the entire population.[186] Nevertheless, several of these policies provide that this priority list can be changed if circumstances so require, thus ensuring that prioritization is no more than a paper promise.[187] This is probably welcomed by people who argue that this administrative prioritization of drinking water eliminates flexibility and 'results in less than optimum utilization of scarce water resources between hydropower and irrigation'.[188] The water policy of the state of Orissa, however, reacts to this trend and specifically provides that any change to the prioritization adopted will require the adoption of a new water policy.[189]

Thirdly, policies emphasize the need for involvement and participation in the planning, design, development and management of water schemes.[190] The inclusion of participatory provisions is meant to benefit users, beneficiaries and other stakeholders.[191] This may appear relatively broad on a first reading but suffers from the same shortcomings highlighted above in the context of the discussion on participation in the context of water sector reforms. The Uttar Pradesh policy sheds interesting light on the substance of participation. It highlights that one of the important measures that need to be adopted is to ensure that local bodies should be involved in the operation, maintenance, and management of water infrastructure with a view to eventually transfer the management to these bodies.[192] This confirms that participation in water policies is not the notion of a right of users of water to participate in the adoption of policies and laws on water, the design of schemes that they are meant to benefit from as well as operational aspects. Additionally, the scope of this limited participation under some policies is narrow. Thus, in Rajasthan, the participatory section of the water policy only addresses irrigation water. Further, participation is restricted to farmers who are users of irrigation water and it covers only the 'management of irrigation systems, particularly in water distribution and collection of water charges'.[193] The idiosyncratic understanding of participation in water policies is confirmed by the fact the Himachal Pradesh water policy lists under different sub-sections of the same

[186] eg National Water Policy 2002, s 8 and Uttar Pradesh Water Policy 1999, s 1(4).

[187] eg Himachal State Water Policy 2005, s 7, Uttar Pradesh Water Policy 1999, s 5(1) and Rajasthan State Water Policy 1999, s 8. Note that Rajasthan State Water Policy (Draft) 2005, s 8 sought to reverse this by ensuring that drinking water remains the top priority even where re-ordering takes place but Rajasthan State Water Policy (Draft) 2008, s 1 does not include this mention anymore.

[188] R Bhatia, 'Water and Energy Interactions' in J Briscoe & RPS Malik, *Handbook of Water Resources in India—Development, Management and Strategies* (New Delhi: The World Bank and Oxford University Press, 2007) 206, 216.

[189] Orissa State Water Policy 2007, s 1(1).

[190] eg Himachal State Water Policy 2005, s 2(1).

[191] eg National Water Policy 2002, s 6(8) and Uttar Pradesh Water Policy 1999, s 14(1).

[192] Uttar Pradesh Water Policy 1999, s 14(1).

[193] eg Rajasthan State Water Policy 1999, s 12. A similar approach is taken by Karnataka State Water Policy 2002, s 6(7).

section on participatory approaches, the participation of local communities and private sector participation.[194]

Fourthly, the devolution of power at the local level is only envisaged in certain specific activities. In a number of areas, the state either seeks to maintain its de facto prerogatives or extend them. Some water policies reassert that the ownership of water resides with the state.[195] Some policies affirm the right of the government to provide for the transfer of water from one river basin to another.[196] This is now being taken up in the context of the river inter-linking scheme.[197] At the state level, an increasing number of states are seeking to control and regulate groundwater whose use has been largely linked to land ownership until now. In other words, decentralization in certain areas is accompanied with attempts at strengthening Union/state government control in other areas.

Fifthly, recent water policies generally promote the use of incentives to ensure that water is used 'more efficiently and productively'.[198] The main consequence which flows from this is the call for private sector involvement in all aspects of water control and use from planning to development and administration of water resources projects.[199] An area which is singled out for private sector participation is urban water supply.[200] Private sector participation is linked to the introduction of water charges for all users as part of a strategy to ensure cost recovery. The basic principle is that water users should pay at least for the operation and maintenance charges linked to the provision of water. One of the consequences is the generalization of meters as well as, in some cases, proposals to introduce mechanisms that automatically cut supply in case of non-payment.[201] In the future, the idea is to move towards full cost recovery which implies that users will have to pay capital costs as well.[202] The draft water policy of Assam specifies that at least 50 per cent of capital costs should be recovered.[203]

Sixthly, some water policies propose the creation of water rights in favour of users.[204] The absence of clear and enforceable water rights is seen as leading to water use inefficiency and conflicts.[205] These rights are seen as a necessary premise for participation in the management of water resources, for the setting up

[194] Himachal Pradesh State Water Policy 2005, s 2(3).
[195] eg Kerala State Water Policy 2007, s 1(2) and Himachal State Water Policy 2005, s 4(5).
[196] eg National Water Policy 2002, s 3(5) and Draft State Water Policy of Assam 2007, s 8(2).
[197] eg Government of India—Ministry of Water Resources, Resolution No.2/21/2002-BM, New Delhi, 13 December 2002.
[198] Maharashtra State Water Policy 2003, s 1(3).
[199] eg National Water Policy 2002, s 13 and Uttar Pradesh Water Policy 1999, s 19(1).
[200] eg Rajasthan State Water Policy 1999, s 9.
[201] eg Draft State Water Policy of Assam 2007, s 8.
[202] eg National Water Policy 2002, s 11.
[203] Draft State Water Policy of Assam 2007, s 9(3).
[204] Uttar Pradesh Water Policy 1999, s 17(1)(d).
[205] Kerala State Water Policy 2007, s 2(2).

of water user associations and for the introduction of trading in entitlements. Trading is in fact specifically proposed in the Maharashtra water policy.[206]

Seventh, policies call for wide-ranging legal and institutional reforms. These include the introduction of various amendments to existing laws as well as the introduction of new laws. Three main aspects are singled out. These are the introduction of a legal framework for the formation of water user associations to decentralize water governance, the introduction of laws providing for the establishment of a water resources authority whose primary characteristic is to be largely independent from existing irrigation and other water resource departments, the creation of water rights in favour of users and the regulation of groundwater.[207]

Eighth, water policies focus on changes to governance in the water sector. New ways to manage water are at the core of most water policies. One of the main changes that are proposed is a move from supply-driven to demand-driven approaches.[208] This is accompanied by an emphasis on the commercialization of water systems. Water policies do not, however, put similar emphasis on broader issues of human rights, environmental sustainability and social equity. Exceptions like the Kerala State Water Policy whose first principle is that water is a human right, show that there exist opportunities for rethinking the development of water policies along different lines.[209] The recent Orissa policy also moves forward in recognizing the need to consider environmental flows in project planning.[210]

b) Water policies in the context of water law

Existing policies must be understood in the context of the place of policies in a law and policy context. A policy in the present context usually refers to 'a course or principle of action adopted or proposed by an organization'.[211] There is thus a large programmatic element in a policy which is supposed to provide a plan of action including objectives and the means to achieve them for what is usually an unlimited period of time. Such policies are by definition non-binding and no enforceable rights ensue.

The implication from the previous elements of definition is that policies like the water policies currently existing in India do not have force of law. In fact, as a rule there is a distinction between policies that are adopted by the government and acts which are adopted by the relevant elected representatives, either MPs or MLAs of the concerned state legislature. Further, a policy can be modified at any time by the government. An act can, however, only be modified in accordance

[206] Maharashtra State Water Policy 2003, s 4(2).
[207] eg Karnataka State Water Policy 2002, s 7, Kerala State Water Policy 2007, ss 2(15) & 2(3) and Uttar Pradesh Water Policy 1999, s 17(1).
[208] eg Himachal State Water Policy 2005, s 11(3).
[209] Kerala State Water Policy 2007, s 1(2).
[210] Orissa State Water Policy 2007, s 7(2).
[211] *OED* (revised edition), C Soanes and A Stevenson (eds) (2005).

with the relevant constitutional procedures. As a result, the legitimacy of legislation is much higher. Further, only acts create rights and obligations.

In principle, the hierarchy between an act and a policy is not a matter for debate. Yet, in the context of water sector reforms, policies are sometimes given much more importance than is warranted. In the context of a new field of regulation, policies can provide an important context that the legislature can use as a basis for the adoption of binding measures. This function of policies is not a matter for concern as long as this does not impede the role and functions of the legislature and the courts. In the context of water sector reforms, this has, however, become a controversial issue because the importance attached to policies has the unwelcome side-effect of indirectly making the role of parliament and legislative assemblies subservient to documents which have no particular constitutional status.

The emphasis put on policies as the main regulatory instrument that provides the framework within which legislatures are supposed to take the binding measures that implement the principles adopted in the policies, is problematic from at least two perspectives. Firstly, the water policies adopted by a number of governments are not instruments which bring together the different strands of existing water law to provide a coherent framework to give more visibility to water law and ensure that the different water laws adopted are coordinated. The water policies that now exist in India rather provide the framework for the overhaul of the water law framework and seek to effect a clean break with past water law. Assuming that legislatures then have to follow all these new principles that they have not contributed to develop, this is damaging to the democratic process. Secondly, the water policies that have been adopted over the past two decades follow a similar scheme. This is easily explained by the fact that all these policies have been and are being adopted as a prelude to the introduction of large-scale water sector reforms and water law reforms. Yet, there are a number of issues that are specific to a particular state. In India, more than in many smaller countries, there is very little in common between the water problems faced by Assam, Himachal Pradesh and Rajasthan. The adoption of state water policies in accordance with the constitutional mandate giving them the main responsibility over water should thus be the place for taking into account these differences in the responses given by different states. These various documents show, however, a lot of similarity with each other. One of the likely reasons for this is the fact that most states have had in-depth interactions with international policy-makers and donors in the field of water. The similarity in water policies thus reflects the engagement of the Union and state governments with international policy makers.[212]

[212] While some commentators refute the idea that the central and state governments may have been influenced by the opinions of the World Bank and other international agencies, Mohile acknowledges, for instance, that '[i]ndirectly, the experience of working with such bodies would have helped in drafting the water policies'. AD Mohile, 'Government Policies and Programmes' in J Briscoe & RPS Malik, *Handbook of Water Resources in India—Development, Management and Strategies* (New Delhi: The World Bank and Oxford University Press, 2007) 10, 18.

The kinds of problems that can arise when policies and acts are intermixed are illustrated by the Maharashtra Water Resources Regulatory Authority Act 2005. This act fails to provide the principles that guide the action of the independent authority. Instead it refers back to the state policy and simply indicates that the 'Authority shall work according to the framework of the State Water Policy'.[213] In the first place this does not pose a problem since the expectation is that the Legislative Assembly adopted the act with the knowledge of the content of the State Water Policy adopted earlier. However, this is problematic because the policy can be changed by the government without referring it to the Legislative Assembly. While a change of policy by the government would not be binding on the Legislative Assembly and could thus be challenged, this has the unwelcome consequence of indirectly giving the government the power to initiate changes that should in principle be undertaken by the legislature until such time as they are struck down by a court. This arrangement is all the more disturbing because the act fails to address the issue of prioritization of water uses. Today, the priority is given to drinking water in the policy but nothing stops the government from amending the policy to read, for instance, like the Himachal Pradesh water policy which provides that the order or priorities can be changed if 'special considerations' warrant such change.[214]

D. Towards Water Law Reforms

Water sector reforms have been implemented in a number of countries over the past couple of decades. An overall assessment of water sector reforms is difficult to make. Indeed, the measure of success and failure depends in part on whether the appraiser agrees with the principles that underlie water sector reforms and in part on whether the reforms have been successful according to their own criteria. As identified earlier in this chapter, the principles that underlie water sector reforms are problematic insofar as they do not provide the basis for a comprehensive reform of the water sector but rather focus on a limited number of factors that sideline, for instance, the human right to water and principles of environmental law.

With regard to an assessment of the reforms from within the framework provided, two different perspectives arise. On the one hand, some reforms, such as participatory irrigation management have been implemented in various countries for a number of years through a variety of schemes. It has been introduced in many regions of the world. On the other hand, water sector reforms have not performed as desired in certain cases. The most prominent of these are failed urban water service privatization schemes, as in the case of the cities of

[213] Maharashtra Water Resources Regulatory Authority Act 2005, s 12(1).
[214] Himachal State Water Policy 2005, s 7.

Cochabamba or Buenos Aires.[215] Additionally, there have been question marks over the sustainability of certain other reforms proposed through project-specific interventions. Thus, in India, analysts have pointed out that efforts at participatory irrigation management schemes have neither shown a potential for replicability nor scaling up.[216]

Controversies over privatization schemes and implementation problems with certain other reforms have led water sector reform proponents to rethink the mode of operationalization. One of the outcomes was a new strategy to promote reforms in a way that would ensure their viability over the long-term even after the specific intervention of a donor through a project terminates. The idea of promoting law reforms as a first or concurrent step to the introduction of project-specific water sector reforms thus gained currency in an attempt to ensure that the failures of the 1990s would not be repeated.

At the same time as a new framework for reforms was being proposed, the regional focus also started shifting. While significant emphasis was put for a number of years on water services reform in Latin America, the past decade has seen a shift to other countries, with India being a key focus of the new path towards reform. Indeed, in the past decade, a host of water projects financed by the World Bank and the Asian Development Bank have been proposed for different aspects of the water sector in different states. The underlying principles that inform the two banks' interventions have not significantly changed. What has changed, however, is the delivery mode. Where water project conditionality was linked to the implementation of the specific project at hand, the new framework is conceived in terms of much broader legislative interventions. One of the most prominent interventions under this new scheme is the introduction of legislation to set up independent water regulatory authorities. This constitutes a broad-ranging and deep reform that does not trigger the kind of immediate changes that project-based interventions can achieve but has the potential to radically transform the water sector for years and decades to come.

The use of water law reforms as a strategy to further the implementation of water sector reforms can contribute to achieving several goals. Firstly, it ensures that the principles for water sector reforms are mainstreamed beyond the physical scope of a specific project and beyond the term of the project. Secondly, it provides the scope for the introduction of reforms, which have a more local tint since they are the initiative of the government or legislature. Thirdly,

[215] On Cochabamba, eg EJ Woodhouse, 'The "Guerra del Agua" and the Cochabamba Concession: Social Risk and Foreign Direct Investment in Public Infrastructure' (2003) 39/2 *Stanford J Intl L* 295. On Buenos Aires, eg A Olleta, 'The World Bank's Influence on Water Privatisation in Argentina—The Experience of the City of Buenos Aires' in P Cullet, A Gowlland-Gualtieri, R Madhav & U Ramanathan (eds), *Water Law at the Crossroads—National and International Perspectives With Special Emphasis on India* (New Delhi: Cambridge University Press, 2009) 230.

[216] N Pant, 'Some Issues in Participatory Irrigation Management' (2008) 48/1 *Economic & Political Weekly* 30.

in the case of socially or politically unpalatable reforms, it provides a way to bring changes in a phased manner.[217]

Water law reforms have been identified as an important component of water sector reforms by institutions promoting reforms. This can be traced over a relatively long period of time. The World Bank thus already recommended in 1998 that revising and strengthening the legislative framework was fundamental to improving conditions in the water sector in general.[218] More recently, the Briscoe-authored World Bank report highlights, among the features of the 'new Indian water state' it calls for, the development of 'a set of laws, policies, capacities, and organizations for defining and delivering an enabling environment, with special emphasis on the establishment and management of water entitlements, and the regulation of services'.[219] Similarly, the Camdessus report identifies an 'inadequate general legal framework' as one of the important governance issues that needs to be addressed.[220] Yet, in this case, the understanding of law reform is much narrower since the Camdessus report only calls for a legal framework that provides adequate incentives to attract finance in the water sector. This does not specifically involve changes to water law but rather to corporate, investment protection, and banking laws.[221]

In practice, ongoing water law reforms are for the most part linked to water sector reforms and based on the same principles. This is, for instance, the case of water user association laws adopted since the mid-1990s as well as laws setting up independent water regulatory authorities. Some of the main reforms, such as the ones just highlighted have been taken up through the adoption of new laws passed by the relevant legislative assembly. These interventions do not, however, correspond to the sum of law reforms that are being introduced. Indeed, a host of other instruments, such as government orders, are also used to set up a new regulatory framework in the water sector. One of the most significant non-legislative interventions was the adoption of the Swajaldhara Guidelines that constituted a framework for completely restructuring drinking water provision in rural areas but were simply adopted in the form of guidelines.[222] The diversity of instruments illustrates the sensitivity of the reforms undertaken. Indeed, while all the above reforms are significant, drinking water reforms are without doubt the most important ones because they concern the most important use of water, that which is covered by the fundamental right to water. The choice of instrument is

[217] cf CL Abernethy, 'Constructing New Institutions for Sharing Water' in BR Bruns, C Ringler & R Meinzen-Dick eds, *Water Rights Reform: Lessons for Institutional Design* (Washington, DC: International Food Policy Research Institute, 2005) 55.

[218] World Bank, India—Water Resources Management Sector Review—Inter-sectoral Water Allocation, Planning and Management (Report No. 18322, 1998) 52.

[219] Briscoe & Malik (n 164 above) 41.

[220] World Panel on Financing Water Infrastructure, Financing Water for All (Marseille: World Water Council, 2003) 9.

[221] ibid 10.

[222] ch 5.B.2.

significant in terms of legitimacy and accountability. It also confirms the lack of pre-eminence given to the human right to water in ongoing reforms.

While most water law reforms are linked to water sector reforms and their principles, exceptions exist. In India, the most significant exceptions are ongoing reforms of groundwater law which follow a different paradigm. This is explained by the fact that the model bill that Indian states are largely following is based on a document first proposed in 1970 which is informed by the policy framework of that era. Recent groundwater laws are not based on a demand-led approach or on cost recovery. Rather, they provide a framework that gives the government a significant role in regulating the use of groundwater. It thus largely runs counter to the other reforms taking place at present. Significantly, new groundwater legislation does not provide a comprehensive response to identified problems because it adopts a technocratic view of groundwater, fails to include social and environmental aspects, fails to make a clean break with land-based rights of access and control that are in large part responsible for the unsustainable use of groundwater and fails to look at surface and groundwater as a unitary resource. At the same time, groundwater law reforms may fit in the context of water sector reforms because the increased government control can be subsequently used to at least partly delink land and water and thus provide the basis for tradable entitlements.

4

Evolving Water Law for the Twenty-First Century

As identified in earlier chapters, there are two different sets of reasons calling for water law reforms. The analysis of existing water law carried out in chapter 2 indicates that there are a number of grounds for introducing changes to existing water laws and introducing new laws since the existing framework is largely outdated and unable to address today's challenges. Chapter 3 has highlighted that law reforms have become one of the elements proposed as part of water sector reforms. These law reforms are on the whole based on the set of principles that inform water sector reforms.

The two kinds of law reforms arising from the analysis carried out in chapters 2 and 3 do not necessarily coincide. This is due to the fact that water sector reforms are not directly informed by the needs or shortcomings of existing water law. Since ongoing reforms are often the by-product of water sector reforms, the analysis carried out in this and the next chapter focuses in large part on a set of reforms that are sectoral in scope and centred around the principles of water sector reforms. The need for a broader framework for law reform that goes beyond existing reforms is then taken up in chapter 6.

Ongoing water law reforms take a number of different forms. The most easily identifiable reforms are legislation in new fields and legislation updating or amending existing acts. This is supplemented by a host of other measures which are sometimes just as significant in their impacts. These include government orders, guidelines and related instruments. Thus, in Maharashtra, the new demand-driven policy for drinking water was in the main introduced through a series of government orders in the early part of this decade.[1] Similarly, at the national level, the radically new set of principles for drinking water introduced earlier this decade came under the guise of the Swajaldhara guidelines.[2]

[1] Project Planning and Implementation Unit of the Government of Maharashtra, Salient Features of Various Government Resolutions Regarding New Demand-Driven Policy, available at <http://mahawssd.gov.in/Homeimage/GIST%20of%20GRS.doc>

[2] On Swajaldhara, ch 5.B. 2–3.

While orders and guidelines are in principle much less important than legislation that benefits from the legitimacy that comes with adoption by the legislature, they are, for instance, central to reform efforts in the case of drinking water.

Water law reforms have a number of different impacts. In certain cases, as in the case of the setting up of new regulatory authorities, the new legislation introduces a completely new governance system whose implications will unravel over many years. In the case of the adoption of groundwater legislation based on the model bill, the reforms that are implemented reflect the thinking of water experts of several decades ago. In yet other cases, the reforms put in place simply formalize existing practices. This is, for instance, the case of the proposed drinking water quality legislation.

This chapter does not seek to analyse all ongoing water law reforms. It examines a limited number of significant changes and focuses specifically on water user association legislation, acts setting up independent water regulatory authorities and groundwater legislation. It does not cover drinking water which is the object of chapter 5.

A. Participation and Decentralization: Water User Association Legislation

Irrigation water is of primary importance in the regulation of water since irrigation uses an overwhelming majority of surface and ground water. Indeed, early water legislation from the late nineteenth century to the middle of the last century put significant emphasis on irrigation. There is thus a long history to irrigation water regulation. Most irrigation legislation needs updating. Indeed, several old colonial acts are still in force today. The need for broad-ranging interventions in irrigation regulation is clearly demonstrated by the fact that existing legislation does not reflect new principles of water regulation and by the fact that irrigation legislation does not, as a rule, recognize the tremendous changes that have taken place with the massive introduction of mechanized devices to pump groundwater over the past five decades.

In reality, the water law reforms that are being introduced do not reflect this need for broad-ranging interventions. They focus on one specific aspect of irrigation, which is the management of water infrastructure by landowning farmers. This is undertaken in the context of the principles of water sector reforms highlighted in chapter 3, with most emphasis given here to the idea of decentralization and participation. The reforms introduced in irrigation management reflect the comments made in the previous chapter concerning the concept of decentralization. While lip service is paid to the constitutionally sanctioned decentralization scheme devolving power to panchayati raj institutions, water user associations (WUAs) are in principle not linked to the panchayati raj institutions.

1. Rationale for reforms

A number of reasons have been offered to justify the need for changes. Firstly, reforms are justified by the need to introduce participation and decentralization. Participation of farmers in the management of irrigation infrastructure is called for with a view to ensure that they have appropriate incentives to maintain canals. One of the main focus areas of participation focuses on operation and maintenance. The need for farmers' organizations to take over operation and maintenance from the government is seen as providing the potential for improving efficiency of management and reducing government costs.[3] This transfer is premised on the shrinking availability of resources for irrigation and the fact that farmers can be more efficient than the government because they are 'unburdened by bureaucratic regulations'.[4]

Participation by farmers and the control they are given over certain management functions is directly linked to the introduction of a duty to take on responsibility for funding. The underlying idea is that where irrigation infrastructure is 'funded by the client', this puts 'the farmer (...) in the driver's seat'.[5] In other words, farmers are now to be considered as 'clients' rather than beneficiaries of government patronage.[6] In the new scheme, a direct link is established between revenue collection and expenditure so that those responsible for operation and maintenance collect and retain water fees.[7]

Participation is linked to decentralization. In principle, this decentralization reflects the 73rd constitutional amendment. In practice, however, this is not the case. Firstly, the reason for decentralization is not necessarily linked to attempts at compliance with the constitutional mandate but has been linked to financial pressure imposed both internally and externally by international lending agencies on governments.[8] Secondly, panchayati raj institutions are said to be unsuitable for taking up the task of decentralization of the management of irrigation. The main arguments revolve around the fact that these bodies have too many other tasks, may lack the expertise to manage water, do not meet often enough to be effective, that not all members of the panchayats are defined as users of irrigation water, that the boundaries of the panchayat do

[3] A Gulati, R Meinzen-Dick & KV Raju, *Institutional Reforms in Indian Irrigation* (New Delhi: Sage, 2005).

[4] ibid 287.

[5] World Bank, India—Water Resources Management Sector Review—Report on the Irrigation Sector (Report No. 18416 IN, 1998) iv.

[6] Gulati, Meinzen-Dick & Raju (n 3 above) 292.

[7] World Bank, India—Water Resources Management Sector Review—Report on the Irrigation Sector (Report No. 18416 IN, 1998) 36.

[8] V Narain, *Institutions, Technology and Water Control* (New Delhi: Orient Longman, 2003) 2.

not necessarily match the relevant hydrological boundaries and that panchayats are mired in politics and factionalism.[9]

This rejection of panchayati raj institutions is problematic and reflects the lack of consideration given to the Indian legal framework in importing participatory irrigation management into India. Indeed, India in common with a variety of federal countries has today a clear framework for administrative decentralization. There must be very strong reasons for bypassing existing decentralized institutions even if, in some states, they do not yet function to full satisfaction. The rejection of existing institutional structures at the local level does not do credit to the proponents of participatory irrigation management. Indeed, in probably all villages, irrigation is an issue which concerns not only landowners but everyone else as well. This is due both to the fact that farmers are not limited to landowners and to the fact that the use of water for irrigation has impacts on water use for domestic water, something that is central to everyone in a given locality. The argument that the panchayat is ill-suited for taking up this task because its boundaries do not correspond to relevant hydrological boundaries is also misplaced.[10]

Secondly, reforms are premised on the need to reduce the role of the government in irrigation. Administrative centralization is seen as a cause of the failure of irrigation schemes because it 'lacks the accountability, corporate management skills and client focus of the private sector, and tends to be remote, top-down and with minimal contact with farmers'.[11] Additionally, in the broader context where government spending is to be reduced, irrigation is one of the major areas targeted for reduced spending.[12] The underlying idea is not for the government to withdraw totally from service delivery but, in particular, to provide water supply at the main delivery points.[13] Overall, irrigation departments are called upon to become smaller entities with their functions reduced to policy formulation, design, investment funding, and legislation.[14]

Thirdly, financial issues play a big role in the proposed reforms. Cost recovery is thus one of the driving principles of ongoing legislative reforms. The rationale for the emphasis on financial sustainability is that this only equates to going back to a situation which prevailed in colonial times and in the early years of independence where water charges were fully covering the costs of maintenance.[15]

[9] Gulati, Meinzen-Dick & Raju (n 3 above) 202 and R Hooja, 'Below The Third Tier: Water Users Associations and Participatory Irrigation Management in India' (2004) 1 *Indian J Federal Studies*, available at <http://www.jamiahamdard.edu/cfs/jour4-1_4.htm>.

[10] ch 6.D.3, p 212.

[11] World Bank, India—Water Resources Management Sector Review—Report on the Irrigation Sector (Report No. 18416 IN, 1998) ii.

[12] S Hodgson, Land and Water—The Rights Interface (Rome: FAO, FAO Legislative Study 84, 2004) 66.

[13] Gulati, Meinzen-Dick & Raju (n 3 above) 290.

[14] World Bank, India—Water Resources Management Sector Review—Report on the Irrigation Sector (Report No. 18416 IN, 1998) v.

[15] Gulati, Meinzen-Dick & Raju (n 3 above) 92.

Yet, unlike in the case of drinking water supply, the proposal here is not to make farmers pay for the whole cost of the irrigation infrastructure.[16] This may sound like a contradiction in terms but is in fact a simple pragmatic realization that while full cost recovery of capital and maintenance costs is the aim of the reforms, its introduction would have far-reaching social consequences, because of the ensuing food price rises.[17] Reforms thus emphasize full cost recovery for operation and maintenance because it is the only realistic option and because governments have usually accorded less importance to maintenance than to the creation of new infrastructure.[18] The justification for cost recovery is that it contributes to reducing inefficient water use. It is acknowledged that this implies much higher water rates, such as the three-fold increase implemented in Andhra Pradesh in 1997.[19]

The focus on cost recovery in the context of surface irrigation is symptomatic of the bias of ongoing reforms. According to water sector reform principles, full cost recovery of maintenance as well as capital costs is the target. Yet, even without looking at the impacts on food prices for the majority of the population of making farmers pay for the building of dams and canals, the simple idea that a nation made of an overwhelming majority of small exploitations could pay for the enormous sums involved in large dams and canal projects is ludicrous. It seems to indicate that reform principles proposed for India are out of tune with the reality on the ground. Nevertheless, these are noteworthy comments because they question the very rationale for large dams feeding large irrigation command areas. If the state is not supposed to invest in them because the benefits are captured by farmers rather than the whole population, this would seem to justify abandoning large schemes altogether. However, as analysed further in the last section of this chapter, ongoing reforms seem to be pulling in two opposing directions at the same time. On the one hand, demand-led management of water infrastructure calls for imposing all direct costs of irrigation on farmers using the water. On the other hand, the Government of India and international lending agencies are again keen to promote large-scale dam building. The two can only be reconciled through a series of convoluted arguments.

Interestingly, regardless of the proposals made, it is small farmers that tend to suffer and be accused of non-performance. Existing reforms are premised in part

[16] Different proposals to charge part of the capital costs to farmers have been made. Planning Commission, Ninth Five Year Plan—Volume 2 (New Delhi: Government of India, 1996) para 4(2) (19) suggested that after covering operation and maintenance costs in full, states might want to consider charging 1% of capital costs to farmers. See also World Bank, India—Water Resources Management Sector Review—Report on the Irrigation Sector (Report No. 18416 IN, 1998) 26 accepting that charging full capital costs 'may not be feasible', but suggesting at least contributions to a 'renewal fund'.

[17] Gulati, Meinzen-Dick & Raju (n 3 above) 113.

[18] Gulati, Meinzen-Dick & Raju (n 3 above) 77.

[19] World Bank, India—Water Resources Management Sector Review—Report on the Irrigation Sector (Report No. 18416 IN, 1998).

on the fact that farmers are accused of using water inefficiently because it is too cheap. At the same time, it is the same farmers that suffer from waterlogging and salinity caused by irrigation.[20] The same reforms whose frontispiece is farmer participation show an uncanny distrust of farmers' local expertise for what is in most cases their sole and unique source of livelihood.

Fourthly, reforms seek to ensure better service to irrigators and better governance of the irrigation infrastructure. This is to be achieved through different strategies. A first approach, derived from the principle of participation, calls for farmers to be involved in the management of irrigation systems. Additionally, farmers are to be seen as clients of a commercial service. In this scheme, they become entitled to put pressure on the service provider to improve performance.[21] A second approach is to foster the involvement of the private sector for a range of activities, including management contracts for irrigation schemes and contracts to perform certain activities under the purview of the irrigation department.[22] Both strategies seek to foster better management of irrigation systems. However, while farmer participation in management and private sector involvement are conceived as two discrete activities, both overlap and may conflict. This is not something that reforms directly address but is an aspect that will require increasing scrutiny since farmer participation conceived as a way to enhance efficiency is likely to end up being more often than not an avenue towards the commercialization of the system rather than an avenue for enhancing the choices and control that farmers have over irrigation systems.

2. Evolution of participatory irrigation management regulation

Irrigation has been a central component of formal water law since the colonial era. The introduction of water laws by the colonial government occurred at a time when it sought to harness the potential of water for productive activities. This focus on water as 'natural resource' to be used for economic development purposes has had lasting impacts and partly explains why the shift to water conceived as a 'commodity' today can go on relatively unchallenged and why the human right to water does not always receive the central position it deserves.

Early irrigation laws attempted to balance the needs of the colonial state to use water for its own needs and the long-standing recognition that water could not be appropriated. The Northern India Canal and Drainage Act 1873 thus entitled the government 'to use and control for public purposes' all surface waters, apart from ponds or other collections of water created by humans.[23] This was reinforced over

[20] eg J-M Faurès, M Svendsen & H Turral, 'Reinventing Irrigation', in D Molden (ed), *Water for Food, Water for Life* (London: Earthscan, 2007) 383.

[21] World Bank, India—Water Resources Management Sector Review—Report on the Irrigation Sector (Report No. 18416 IN, 1998) iv.

[22] ibid v.

[23] Northern India Canal and Drainage Act 1873, preamble.

time and later irrigation acts, such as the Madhya Pradesh Irrigation Act 1931, asserted that rights in all surface water vest in the government.[24] The main result of the regulatory framework was thus to extend government control over irrigation water and by extension most surface water.

This new regulatory framework vesting increasing rights in the government had the correlative impact of restricting or extinguishing traditional rights enjoyed by individuals and communities over water for irrigation or other purposes. Existing scholarship does not provide a comprehensive picture of the customary rights that individuals and communities enjoyed but available evidence shows that there were well-established and functioning systems in place in many parts of the country and that in the absence of centralized control, irrigation was managed at the local level by beneficiary communities.[25]

While irrigation laws on the whole significantly strengthened the power of the government, there was a realization even by the colonial government that a completely centralized system would not work efficiently. The talk was not necessarily about giving rights to irrigators but there was at least a recognition of the need to involve them. For instance, both the first irrigation commission (1901–1903) and the royal commission on agriculture (1928) recommended participatory management in irrigation.[26] Additionally, the Madhya Pradesh Irrigation Act 1931 provided for the setting up of irrigation panchayats for every village.[27] This scheme did not correspond to a democratic process of decentralization since the establishment of the irrigation panchayat was itself at the discretion of the collector. Additionally, the irrigation panchayats were conceived with very limited autonomy and rather tasked on the whole with assisting government officers in their work. This constituted, nevertheless, a recognition that complete centralization was not an appropriate strategy.

Changes with regard to the devolution of irrigation structures to irrigators over the past few decades can be divided into two separate components. On the one hand, different experiments have been carried out since the mid-1970s, for instance, through the implementation of pilot projects. These became progressively more organized and systematic and for the past decade participatory irrigation management has been a core element of attempts to restructure the irrigation sector.[28] On the other hand, the regulatory framework started evolving later. The relatively recent Bihar Irrigation Act 1997 is, for instance, largely premised on old principles highlighted above. Similarly, the model irrigation bill proposed by the India Law Institute in the 1970s suggested the formation of

[24] Madhya Pradesh Irrigation Act 1931, s 26.

[25] eg D Mosse, *The Rule of Water—Statecraft, Ecology and Collective Action in South India* (New Delhi: Oxford University Press, 2003).

[26] A Narayanamoorthy & RS Deshpande, *Where Water Seeps!—Towards a New Phase in India's Irrigation Reforms* (New Delhi: Academic Foundation, 2005) 201.

[27] Madhya Pradesh Irrigation Act 1931, s 62.

[28] N Pant, 'Some Issues in Participatory Irrigation Management' (2008) 48/1 *Economic & Political Weekly* 30.

committees where the government found it suitable and under conditions that put water committees at the mercy of the government, for instance, by providing that members would hold office 'at the pleasure' of the government.[29]

Since the early 1990s, the regulatory framework for participation by irrigators in the management of irrigation has completely changed. This is due to two parallel, though largely unconnected developments. Firstly, the adoption of the 73rd amendment to the Constitution provided a new legal basis for reversing the trend towards centralization in irrigation regulation by specifically seeking to endow panchayats with control over minor irrigation schemes as well as more general water management and watershed development.[30] This appears to be an apt and sensible response to the perceived widespread failure of centralized systems to deliver irrigation water at the right time and in appropriate quantities to all farmers. Secondly, the worldwide policy push for the setting up of water user associations was also taken up in India in earnest from the early 1990s. This is visible in the context of domestic policy documents as well as World Bank projects and policy proposals.[31] The recommendations put forward included the setting up of water user associations but emphasized mostly the issue of transferring responsibility of recovery of operation and maintenance costs to beneficiaries.[32]

Since the mid 1990s, there has been a dramatic increase in the rate of change in various states. While forms of participation had been proposed throughout the twentieth century, more widespread changes started with the emergence of donor funding focusing on the setting up of water user associations.[33] The changes that have taken place in practice do not follow the lead of the 73rd constitutional amendment seeking democratic decentralization of irrigation governance but rather the model seeking to devolve certain rights and obligations concerning the management of irrigation systems, in particular financial aspects.

3. The proposed reform model

Over the past two decades, a number of new measures have been introduced in the regulation of irrigation. These new measures have focused on irrigation management and the main response to perceived inadequacies has been the proposal for the compulsory setting up of water user associations. Two main characteristics of the present reforms stand out at the outset. Firstly, ongoing reforms are not reforms that comprehensively address all issues related to

[29] Indian Law Institute, Model Irrigation Bill 1977, s 45(5).
[30] Constitution, art 243G and Schedule XI, s 3.
[31] Narayanamoorthy & Deshpande (n 26 above) 203.
[32] ibid 203.
[33] A Sekhar, 'Development and Management Policies—Perspective of the Planning Commission' in J Briscoe & RPS Malik, *Handbook of Water Resources in India—Development, Management and Strategies* (New Delhi: The World Bank and Oxford University Press, 2007) 47 and Pant (n 28 above) 30.

irrigation. The reforms pick up a limited number of issues that are addressed through the setting up of new institutional frameworks. In fact, comprehensive reforms of irrigation laws are called for and should be undertaken as a matter of priority. This is, for instance, due to the fact that a number of irrigation acts based on outdated principles are still in force. It is thus surprising to note that the reforms proposed and being implemented take a narrow view of the need for change. Secondly, ongoing reforms follow a pattern. Despite the relatively frequent reassertion that a single model is not suitable for a diverse country like India, the acts introduced over the past decade propose participatory irrigation management reforms that are similar in different parts of the country. This cannot be explained by the similar conditions that different states face given the diversity of climatic conditions between Rajasthan and Orissa, for instance.[34] The similarity of framework that can be identified between the different acts implies that there are a number of principles and assumptions that are common to most of them. Interestingly, this finding is also valid in international comparison. Indeed, an FAO study of water user association legislation found that 'the basic principles on which [water user organizations] operate, as well as the legal rules that underpin those principles, are surprisingly similar'.[35]

The primary rationale for water user associations is to transfer to actual irrigators the management of the irrigation infrastructure they directly benefit from. This transfer from irrigation departments to farmers is justified by the principles of decentralization and participation. The meanings given to these two notions correspond closely to the analysis carried out in chapter 3. Decentralization is not used in its usual meaning but serves as the background for the setting up of new local bodies that are outside of the panchayat system. As a result, decentralization envisaged under water user associations does not include universal membership but only landowners and land occupiers.[36] Additionally, water user associations, unlike panchayati raj institutions, only exceptionally include reservation for women or scheduled castes and scheduled tribes.[37]

With regard to participation, the understanding under water user association acts is limited to specific water management tasks. This is, for instance, the case in Maharashtra where the two main objects of water user associations are to ensure an equitable distribution of water among its members, to adequately maintain irrigation systems and to ensure efficient, economical and equitable distribution and utilization of water to optimize agricultural production.[38]

[34] Orissa Pani Panchayat Act 2002 and Rajasthan Farmers' Participation in Management of Irrigation Systems Act 2000. Note eg ss 16 of both acts concerning the 'objects' of farmer organisations, which are virtually the same.

[35] S Hodgson, Legislation on Water Users' Organizations—A Comparative Analysis (Rome: FAO, FAO Legislative Study 79, 2003) 2.

[36] eg Orissa Pani Panchayat Act 2002, s 3(4).

[37] Chhattisgarh sinchai prabandhan me krishkon ki bhagidari adhiniyam 2006, s 5.

[38] Maharashtra Management of Irrigation Systems by Farmers Act 2005, s 4(1).

The emphasis on the devolution of operation and maintenance to the new associations implies that financial aspects are given a lot of attention. Thus, water user associations usually have to be financially independent and therefore need to receive an income that is sufficient to allow them not to go bankrupt.

Maharashtra stands out among the various states where participatory irrigation management legislation has been enacted over the past decade. Firstly, participatory irrigation management has been discussed and implemented in selected areas in Maharashtra since the 1970s.[39] Additionally, participatory irrigation management guidelines were put in place in the early 1990s.[40] Secondly, the Maharashtra Management of Irrigation Systems by Farmers Act 2005 (MISFA) is not only a recent act but also the most evolved in the sense that it is linked to other institutional reforms adopted under the Maharashtra Water Resources Regulatory Authority Act 2005. Indeed, while the MISFA provides a decentralization scheme towards farmer involvement in irrigation at the local level, it also gives significant powers to the Maharashtra Water Resources Regulatory Authority or other designated authorities. In particular, they have the power to determine the command area of an irrigation project for which a WUA must be constituted. Further, the same authority can also amalgamate or divide existing WUAs on a hydraulic basis and 'having regard to the administrative convenience'.[41] In other words, the power granted at the local level is limited by the fact that authorities have the largely discretionary power to make and break WUAs.

The system set up under the act is constraining insofar as once a WUA has been set up, no water is supplied to anyone individually outside the WUA framework and the scheme is binding on all land holders and occupiers. In this sense, WUAs are forced to take on the burden of administering the irrigation system and are largely left to sort out ways in which they want to achieve this. Further, the act provides a uniform model of WUAs regardless of existing arrangements at the local level and regardless of their success at equitably and sustainably using water.

The framework provided under the act seeks to balance benefits and burdens. On the one hand, WUAs are meant to benefit from a more assured water supply and more control over water allocated to them. Further, it is the Authority's duty to supply the amount of water they are entitled to receive. They also have the right to use groundwater in their command area on top of the entitlement they receive from canals. On the other hand, the act gives WUAs a number of powers which are in fact responsibilities. This includes a number of functions which include the

[39] J McKay & GB Keremane, 'Farmers' Perception on Self Created Water Management Rules in a Pioneer Scheme: The Mula Irrigation Scheme, India' (2006) 20 *Irrigation & Drainage Systems* 205, 207 and N Pant, 'Impact of Irrigation Management Transfer in Maharashtra—An Assessment' (1999) 34/13 *Economic & Political Weekly* A-17.

[40] Cooperative Water Users' Association Guidelines 1994, summarized in V Narain, *Institutions, Technology and Water Control* (New Delhi: Orient Longman, 2003) 82.

[41] Maharashtra Management of Irrigation Systems by Farmers Act 2005, s 5.

regulation and monitoring of water distribution among WUA members to the assessment of members' water shares, the responsibility to supply water equitably to members, the collection of service charges and water charges, the carrying out of maintenance and repairs to the canal system and the resolution of disputes among members.[42] These are extensive and possibly burdensome powers. WUAs are not only given the task to manage the infrastructure but also to provide an institutional structure that equitably provides all the services that a public authority would provide. While such arrangements would be an appropriate choice if WUAs were linked to panchayati raj institutions, it is difficult to see how an association of landholders that lacks in democratic legitimacy can perform all these tasks in an equitable and sustainable manner for its members and for the broader society around it. An attempt is made to include women and SC/STs in the Rules but these provisions remain weaker than in the case of panchayati raj institutions.[43] In other words, the existing legislation is both onerous on WUAs who seem to be saddled with more responsibilities than rights and is at the same time unlikely to provide a framework leading to a more socially equitable access to and sharing of water.

The section concerning the powers and responsibilities of WUAs is complemented by a section concerning financial arrangements. As specified under Section 54, the main sources of funding for WUAs will not come from the government. WUAs are meant to meet their expenses from the proceeds of water charges, borrowings, donations, and grants. In other words, the act seeks to ensure that WUAs are financially independent and financially viable, a fact which is confirmed by the encouragement given to WUAs to engage in additional remunerative activities, including the distribution of seeds, fertilizers, and pesticides or marketing of agricultural produce which are only indirectly related to irrigation.[44]

The impacts of the MISFA in practice cannot yet be comprehensively evaluated. Indeed, only a small proportion of existing associations have been reconstituted under the new act and it is thus difficult to identify differences generated by the new regime. There are other WUA acts that have been in place for a longer period of time. This is, for instance, the case of Andhra Pradesh which was an early adopter of water law reforms. In principle, the Andhra Pradesh Farmers' Management of Irrigation Systems Act 1997 should provide lessons of more than a decade of implementation. Yet, the experience remains relatively limited. This seems partly due to the fact that WUAs have not done much more than conduct meetings.[45] This is related to the unwillingness of the irrigation department to

[42] Maharashtra Management of Irrigation Systems by Farmers Act 2005, s 52.

[43] Maharashtra Management of Irrigation Systems by Farmers Rules 2006, ss 6–7 make an attempt at integrating women in the managing committees but the provisions concerning SC/STs are much weaker than under the panchayati raj institutions.

[44] Maharashtra Management of Irrigation Systems by Farmers Act 2005, s 4(2).

[45] V Ratna Reddy & P Prudhvikar Reddy, 'How Participatory is Participatory Irrigation Management?—Water Users' Associations in Andhra Pradesh' (2005) 40/53 *Economic & Political Weekly* 5587, 5589.

devolve its powers to WUAs.[46] Further, while WUAs were given full support by the former chief minister in the late 1990s for political reasons, little was done subsequently to make them effective on the ground.

4. Assessing participatory irrigation management laws

The model being implemented in the states that have adopted WUA legislation and in other states where WUAs are being implemented through other schemes is the translation of what is seen as a universally valid framework. The assumption that a single model can work everywhere is problematic because of the diversity of situations that exist. Indeed, it is surprising that WUA legislation which is premised on the need to give farmers a more important role in irrigation does not build on existing schemes or on older effective systems.

WUA legislation raises a number of additional concerns. Firstly, WUA legislation is supposed to give farmers much greater control over irrigation systems and introduce new rights for irrigators. While a certain number of usufructuary rights are defined, these come with two major caveats. Rights conceded in new legislation are balanced with a set of obligations which make the exercise of these rights difficult and may vitiate the gains obtained by farmers in the process. Additionally, the way in which irrigators' rights are defined usually gives the WUA a right to receive water in bulk from the irrigation department but fails to define the consequences arising from a breach of this right.[47] The broader issue concerning rights is the absence of transfer of ownership. In fact, while several acts talk about the need to inculcate farmers with 'a sense of ownership of the irrigation system',[48] this is largely in a manner of speech. Indeed, the MISFA specifically confirms that the ownership of a canal system whose management is handed over to a WUA remains with the government.[49] The absence of effective control by the WUA is confirmed by the fact that the government can take control back in the larger public interest.[50] The overall impact is that a system which is meant to strengthen farmers' control over irrigation systems seems to impose more obligations than rights. In other words, the balance between the obligations imposed on farmers that take over the management of irrigation systems and the actual rights granted to them is not in their favour and diverges

[46] ibid 5594.

[47] V Upadhyay, 'Canal Irrigation, Water User Associations and Law in India—Emerging Trends in Rights-based Perspective' in P Cullet, A Gowlland-Gualtieri, R Madhav & U Ramanathan (eds), *Water Law at the Crossroads—National and International Perspectives With Special Emphasis on India* (New Delhi: Cambridge University Press, 2009) 110.

[48] Orissa Pani Panchayat Act 2002, s 16, Madhya Pradesh sinchai prabhandan me krishakon ki bhagidari adhiniyam 1999, s 16, Chhattisgarh sinchai prabandhan me krishkon ki bhagidari adhiniyam 2006, s 24 and Maharashtra Management of Irrigation Systems by the Farmers Act 2005, s 4(1)(v).

[49] Maharashtra Management of Irrigation Systems by the Farmers Act 2005, s 69.

[50] ibid.

from the expectations raised by the reforms. The same problem has been identified with regard to participation that is found to be lacking in practice because the process is still mostly directed from the top.[51]

Secondly, WUA legislation is in general exclusionary insofar as it restricts membership of WUAs to land holders or occupiers.[52] While WUAs are premised on their capacity to foster more equitable sharing of water among users, the notion of equity promoted through WUAs is one that is group-specific. It is inappropriate to restrict membership of WUAs to land occupiers since water is a common substance used by all inhabitants of a given locality. Additionally, the restriction to land holders reinforces existing social and economic inequalities since it is often the rich, who are also often from higher castes, who own land. A broader definition of water users can be envisaged and, in fact, the Madhya Pradesh legislation extends membership to anyone who uses 'water for agriculture, domestic, power, non-domestic, commercial, industrial or any other purpose from a Government source of irrigation'.[53]

Thirdly, WUA legislation raises questions concerning equity and accountability. Indeed, in practice, it has been found that the participatory rationale fails to do more than displace problems from a higher to a lower level. Thus, it was, for instance, found in Andhra Pradesh that 88 per cent of canal WUA presidents were upper caste farmers.[54] Additionally, only 1.2 per cent of presidents were women who constituted only 0.1 per cent of members of canal WUAs overall.[55] The extremely skewed gender representation is not a particularly surprising finding. It is, however, problematic in the aftermath of the adoption of the 73rd amendment to the Constitution that specifically enshrined the need for reservation for SC/STs and for women.[56] There is nothing to indicate in theory or practice that WUAs would be immune to the kind of problems that make reservation a necessity for panchayati raj institutions.[57] Since WUAs are premised on their democratic nature linked to the emphasis on participation, the rejection of the panchayati raj model seems at best unwelcome. While the panchayat system may not be an ideal framework, it is a comprehensive framework that seeks to foster effective decentralization and democracy at the local level. Existing WUA legislation has no contribution to make in this area and needs comprehensive rethinking.

[51] Narayanamoorthy & Deshpande (n 26 above) 270.

[52] eg Orissa Pani Panchayat Act 2002, s 3(4), Rajasthan Farmers Participation in Management of Irrigation Systems Act 2000, s 4(2) and Maharashtra Management of Irrigation Systems by the Farmers Act 2005, s 2(1)(w).

[53] Madhya Pradesh sinchai prabhandan me krishakon ki bhagidari adhiniyam 1999, s 2(1)(r). Note that the state of Chhattisgarh carved out of Madhya Pradesh has retained exactly the same definition in its 2006 legislation. Chhattisgarh sinchai prabandhan me krishkon ki bhagidari adhiniyam 2006, s 2(1)(z)(b).

[54] Ratna Reddy & Prudhvikar Reddy (n 45 above) 5588.

[55] ibid.

[56] Constitution, art 243(D).

[57] As noted at n 37 above, this is confirmed by Chhattisgarh legislation which includes reservation.

Another concern raised by WUA legislation relates to the accountability framework. As indicated earlier, one of the premises for the introduction of WUA legislation is the lack of accountability of the government in delivering irrigation benefits to farmers. Yet, in the process of taking this responsibility away from the government, reforms impose new burdens on farmers and fail to ensure that disputes are solved in an equitable manner. Indeed, the dispute settlement framework proposed puts responsibility for resolving disputes in the hands of WUAs.[58] This is welcome as a step towards decentralized justice where simple social pressure can be used.[59] However, since WUAs are not perfect social institutions, a framework to ensure just and equitable decisions must be imposed to ensure that dispute settlement does not result systematically in decisions that favour the richer and more powerful members. Besides the absence of a framework guiding dispute settlement, WUA legislation fails to specifically provide for an appeal through the regular court system.[60]

Overall, the introduction of WUAs is based on principles that meet with widespread approval. Thus, the need for effective participation by farmers or the need for decentralization according to the principle of subsidiarity are supported by most people. Yet, WUA acts are controversial because they are based on the specific understanding of these principles in the context of water sector reforms. Thus, by sidelining constitutionally-sanctioned democratic principles of decentralization without putting forward an alternative framework to foster equity and democracy, WUA legislation provides a framework based on a narrow vision of the needs of the sector. The deeper problem is that WUA legislation assumes that a set of principles focusing on the management of water infrastructure will in itself solve all other problems. Not only is this impossible in theory or practice but this also sidelines the need for broader reforms of irrigation laws that need to be comprehensively revisited. The broader reforms that are needed concern not only the need to address managerial and economic issues but all other relevant issues, as well as the recognition that water for irrigation is not distinct from water for drinking and other uses.

B. Institutional Reforms: Precursor for Further Changes

Institutional changes proposed through water sector reforms are some of the broadest interventions being introduced. The most important of these is the setting up of independent water regulatory authorities. Their significance is twofold. These

[58] eg Chhattisgarh sinchai prabandhan me krishkon ki bhagidari adhiniyam 2006, s 43, Tamil Nadu Farmers' Management of Irrigation Systems Act 2000, s 36 and Maharashtra Management of Irrigation Systems by the Farmers Act 2005, s 63.

[59] In the Mahatma Jotirao Phule association, Ozar, Nasik district, for instance, a farmer breaking rules will first be named in a public meeting. In a second stage, if she fails to amend, a fine and stronger social pressure is applied.

[60] eg Chhattisgarh sinchai prabandhan me krishkon ki bhagidari adhiniyam 2006, s 44 and Tamil Nadu Farmers' Management of Irrigation Systems Act 2000, s 37.

new authorities directly contribute to a comprehensive refashioning of the institutional landscape in the water sector. Additionally, they also constitute an intermediary step towards the introduction of further reforms at a later stage within the context provided by these new institutions. Thus, while privatization is not a direct agenda of the regulatory authorities set up to-date, the framework which is put in place will make the introduction of certain forms of privatization easier in the future.

1. Emphasis on regulation

As indicated in chapter 3, changes in the governance of the water sector have been central to water sector reforms. Different types of changes have been proposed over time. In the case of an early reform effort started in Chennai about three decades ago, the emphasis has been on changing the principles on which the existing institution functions. Thus, over the past couple of decades Metrowater has been increasingly turned into a focused institution with a commercial outlook on its mandate.[61]

Since the early 1990s, policy debates have moved towards suggesting the need for entirely new institutions that take over functions previously performed by government departments. This process has been directly linked to moves towards the privatization of water services. This is significant because the new 'regulatory' institutions set up in the case of privatization are meant to provide a bridge between the newly inducted private sector actors and the government that has had to relinquish its existing powers. The primary feature of these regulatory bodies is to be independent from the government. Their success is premised on freedom from political interference together with a strong culture of promotion of the public interest.[62] Regulatory authorities are also supposed to investigate companies on performance benchmarks, share information with the public and foster public participation.[63]

The Water Services Regulation Authority in England and Wales that was set up as the economic regulator taking over the functions of the Director General of Water Services is tasked, for instance, with setting water prices, promoting competition and protecting consumers' interests.[64] It is structured in the same way as other economic regulators in the UK. The latter feature indicates that water is considered like any other privatized product or service. This is of concern because there is, in fact, little in common between water and electricity. The latter

[61] K Coelho, 'The Slow Road to the Private—A Case Study of Neo-Liberal Water Reforms in Chennai' in P Cullet, A Gowlland-Gualtieri, R Madhav & U Ramanathan (eds), *Water Law at the Crossroads—National and International Perspectives With Special Emphasis on India* (New Delhi: Cambridge University Press, 2009) 81.

[62] United Nations Development Programme, *Human Development Report 2006—Beyond Scarcity: Power, Poverty and the Global Water Crisis* (New York: UNDP, 2006) 100.

[63] ibid 100–1.

[64] Water Industry Act 1991.

comparison is especially relevant in India where the regulatory commissions set up in the electricity context provided the broad model for the new water regulatory authorities.[65]

Even though there is little to justify thinking about the institutional framework for water along the lines of something as different as electricity, this is indeed what was encouraged. Thus, more than a decade ago, the World Bank was already advocating, in the context of the opening up of irrigation to private sector investment, the establishment of independent regulatory entities to regulate prices and costs.[66] The two entities that were mentioned as models were Orissa's power sector and England and Wales' water sector.[67]

While the conception of regulatory entities modelled on sectors that share little with water seems inappropriate, this does not mean that the proposition for new institutions in the water sector is unwelcome per se. In fact, some of the problems that exist at present cannot be effectively addressed because of the lack of institutions that can have an overall perspective on water. Thus, in a context where the simple question of use of water is compounded with increasingly significant questions related to water quality, the traditional distinctions between surface and ground water or between irrigation and drinking water are increasingly irrelevant. Institutions that can address these challenges are necessary. The new regulatory authorities are, however, not the forums that will be able to effectively address these challenges because they have not been conceived for this.

2. New water institutions

The setting up of entirely new regulatory entities was first taken up in Andhra Pradesh in the mid-1990s when Andhra found itself an early adopter of a number of reforms. This included significant reforms that were cautioned and fostered by the World Bank.[68] Andhra Pradesh was also an early adopter of water law reforms. In particular, it adopted an act to set up a new water regulatory entity. The Andhra Pradesh Water Resources Development Corporation was created under the act by the same name in 1997.[69]

Early reforms started in Andhra Pradesh were not immediately followed in other states. This was partly linked to a political situation that was not conducive to such sweeping reforms in other states. Yet, progressively a number of other

[65] cf NK Dubash, 'Independent Regulatory Agencies: A Theoretical Review with Reference to Electricity and Water in India' (2008) 43/40 *Economic & Political Weekly* 43.

[66] World Bank, India—Water Resources Management Sector Review—Report on the Irrigation Sector (Report No. 18416 IN, 1998) v.

[67] ibid.

[68] eg World Bank, Third Andhra Pradesh Irrigation Project (Staff Appraisal Report No. 16336-IN, 1997) and World Bank, Andhra Pradesh Economic Restructuring Project (Project Agreement, Credit Number 3103 IN, 1999).

[69] Andhra Pradesh Water Resources Development Corporation Act 1997.

states started considering institutional reforms. In most cases, reform efforts have been linked either directly or indirectly to World Bank loans. This is, for instance, the case in Delhi, Madhya Pradesh, Maharashtra and Uttar Pradesh. In Delhi, reforms were proposed through the Delhi Water Supply and Sewerage project of the World Bank that should have been well under way now.[70] However, a strong public campaign forced the Government of Delhi on the defensive and the project has been on hold since late 2005.[71] One of the changes that was being proposed alongside the project was the establishment of an 'apolitical' Water and Waste-water Regulatory Commission that would have led to the restructuring of the Delhi Jal Board (DJB) either by setting up a private company taking over the whole DJB and therefore becoming a monopoly supplier or by setting up several companies taking up separate tasks or separate areas of Delhi.[72] In Madhya Pradesh, a project that is ongoing until 2011 imposed the drafting of legislation for the establishment of an autonomous State Water Regulatory Tariff Commission by the end of 2005.[73] It is understood that this legislation has been drafted but the draft has neither been passed nor been made available to the public.

To-date the most sweeping institutional reforms have been introduced in Maharashtra, which adopted the Maharashtra Water Resources Regulatory Authority (MWRRA) Act 2005.[74] Arunachal Pradesh has followed suit by adopting exactly the same legislation.[75] Uttar Pradesh which was supposed to establish a State Water Tariff Regulatory Commission by 2003 as part of conditionality imposed by the World Bank, has now adopted the Uttar Pradesh Water Management and Regulatory Commission Bill 2008 which is substantially based on the MWRRA model.[76] Similar legislation can be expected in more states as the Planning Commission has called on all states to set up regulators based on the Maharashtra model.[77] While domestic institutions make their own call for reform, the adoption of the MWRRA Act is first of all related to World Bank policy prescriptions. While the adoption of the act is not part of project conditionality, it was adopted just before the signing of a major World Bank project and the project document lauds the Government of Maharashtra for having 'taken a number of bold and path-breaking actions' between 2003 and 2005, a period during which the Bank acknowledges that it was 'a critical knowledge/

[70] eg World Bank, Delhi Water Supply and Sewerage Project (Project Information Document, Concept Stage 31803, 2005).

[71] A Bhaduri & A Kejriwal, 'Urban Water Supply: Reforming the Reformers' (2005) 40/53 *Economic & Political Weekly* 5543.

[72] Delhi Water and Wastewater Reforms Bill 2004, s 17.

[73] World Bank, Madhya Pradesh Water Sector Restructuring Project (Project Appraisal Document, Report No. 28560-IN, 2004) 10.

[74] Maharashtra Water Resources Regulatory Authority Act 2005.

[75] Arunachal Pradesh Water Resources Management Authority Act 2006.

[76] Concerning World Bank conditions, see World Bank, Uttar Pradesh Water Sector Restructuring Project (Project Appraisal Document, Report No. 23205-IN, 2001) 35.

[77] Planning Commission, Mid-Term Appraisal of the Tenth Five Year Plan (2005) 226.

advocacy partner to the state'.[78] Additionally, the specific conditions of the project include the establishment of the MWRRA by 31 December 2005 and its operationalization by 30 September 2006.[79]

The basic rationale for the MWRRA is the same as the Andhra Pradesh Corporation. Yet, in the intervening eight years, the framework for independent regulation had evolved. Indeed, the MWRRA is considered sufficiently different—in reforms terms sufficiently more progressive—for it to become the model that even an already reformed state like Andhra Pradesh wants to follow.[80] One of the changes that has taken place is that the MWRRA is conceived as being more 'independent' because its membership does not directly include any incumbent minister or civil servant.[81] This difference may be more theoretical than real to the extent that the chairperson of the MWRRA must statutorily be a former top civil servant and that the current selection procedure favours the choice of civil servants for the other members.[82] Yet, it goes one step further towards progressively taking significant powers related to water away from the government. Indeed, the basic rationale of the MWRRA Act is to restrict the state's influence in water allocation and use. In other words, the water regulatory authority is to take on some of the functions currently performed by irrigation departments and other water-related government institutions.

The establishment of the MWRRA will have significant impacts in the water sector that go much beyond the specific changes in the institutional framework. Thus, the mandate of the MWRRA is deeply influenced by the logic of integrated water resources management. Planning for the use of water resources is to be undertaken at the river basin level. In consideration of the broad scope of planning envisaged for the MWRRA, it has been given wide-ranging powers:[83]

- Its first broad prerogative is to establish a regulatory system for the water resources of the state, including surface and ground waters, to regulate their use and apportion entitlements to use water between different recognized categories of use;
- Concurrently, it must promote the 'efficient' use of water, minimize wastage and fix 'reasonable' use criteria;[84]

[78] World Bank, Maharashtra Water Sector Improvement Project (Project Appraisal Document, Report No. 31997-IN, 2005) 1.
[79] ibid 9.
[80] Address by SP Thakkar, Principal Secretary, Irrigation and Command Area Development, Andhra Pradesh Government at the IWMI-Tata Water Policy Research Programme, 7th Annual Partners' Meet, ICRISAT Campus, Patancheru, 3 April 2008.
[81] Maharashtra Water Resources Regulatory Authority Act 2005, s 4.
[82] ibid s 4(1).
[83] ibid s 11.
[84] Uttar Pradesh Water Management and Regulatory Commission Bill 2008, s 12(k) is even more direct and provides that the Commission must promote 'competition, efficiency and economy' in the water sector.

- The MWRRA also has the task of allocating specific amounts to specific users or groups of users according to the availability of water;
- It is further required to establish a water tariff system as well as to fix the criteria for water charges. This is to be done on the basis of the principle of full cost recovery of management, administration, operation, and maintenance of irrigation projects;
- One of the important tasks entrusted to the MWRRA concerns its role in laying down criteria for the issuance of water entitlements. According to Section 11(g)(ii), criteria are to be laid out for the issuance of bulk water entitlements for all the main uses of water including irrigation, rural and, municipal water supply as well as industrial water supply. The MWRRA has significant latitude in determining priorities of use among the main uses since the act does not include specific guidelines;
- Another task assigned to the MWRRA is the setting up of criteria for trading in water entitlements or quotas. Since the very idea of trading in water entitlements is novel, the act specifically indicates that the premise for trading is that entitlements 'are deemed to be usufructuary rights which may be transferred, bartered, bought or sold on an annual or seasonal basis within a market system and as regulated and controlled by the Authority'.[85]

The extensive powers of the authority are to be exercised within the framework of the state water policy and additional principles found in the act.[86] Some specific features can be highlighted. Firstly, with regard to water quality issues, the authority has to work on the basis of the polluter pays principle. Secondly, the volumetric amount of water made available to holders of water entitlements is to be fixed according to specific criteria. These include, for instance, the need for equitable distribution of water between all land holders and the grandfathering of existing private sector lift irrigation schemes for five years. Thirdly, any person with more than two children has to pay 50 per cent more than the prevailing rates to get entitlement of water for agriculture. These three different elements indicate the breadth of factors that the authority has to take into account.

Another characteristic of these guiding policies is that they have the potential to conflict with each other. Thus a small landowner with three children may have to pay 50 per cent more for his/her water than a neighbouring big farmer even though the principle of 'equitable distribution' should be understood as giving priority to meeting the water needs of small and poor farmers. Additionally, the principle of 'equitable distribution' only seems to apply between land occupiers which implies that anyone not occupying any land does not fall within the purview of this provision.

One of the important consequences of the setting up of a water regulatory authority relates to the strengthened control over water resources which is

[85] Maharashtra Water Resources Regulatory Authority Act 2005, s 11(i)(i).
[86] ibid s 12.

proposed. The Act provides as a general principle that any water from any source can only be used after obtaining an entitlement from the respective river basin agency.[87] This is qualified by a few exceptions such as wells (including bore and tube wells) used for domestic purposes or the grandfathering of existing uses of water for agriculture, at least in an initial phase.

The tension between the desire to steer the management of water resources away from the government and the need for a strong regulatory framework to enforce the new measures proposed is also apparent with regard to the emphasis on 'equity'. In principle, the mandate of the MWRRA includes an interesting objective of ensuring equitable allocation and use of water. Yet, in reality, the MWRRA is neither mandated to take a broad view of equity, nor would it be desirable for this body to do so. What the MWRRA is mostly tasked to do is ensure equity among people defined as water users. In the case of irrigation, for instance, this means ensuring equity in allocation among landed farmers,[88] but does not include equity issues concerning the sharing of water between landed and landless farmers or between farmers situated in areas where surface water irrigation exists and in the majority of areas where it does not. Additionally, the notion of equity to be used by the MWRRA is tainted by the fact that it must at the same time ensure that water charges reflect the full recovery of the cost of the irrigation management, administration, operation and maintenance of water resources projects.[89]

The broader limitation of the MWRRA and of similar authorities is that they do not have the democratic legitimacy to engage in some of the broad policy decisions that need to be taken. On the one hand, it is appropriate that the MWRRA should not be tasked with the implementation of the human right to water or of setting out environmental requirements for water regulation since it neither has the legitimacy nor the expertise to do so.[90] On the other hand, the MWRRA appears as the de facto comprehensive regulator for all water uses and may indeed end up having a direct and/or indirect influence on all water uses. This is problematic from an institutional point of view because while the MWRRA is subject to government control, the very point of setting up regulatory authorities is to give them more independence from the government than the institutions they replace.[91] This independence is meant to ensure that the authority better represents the interests of the specific stakeholders that are recognized, such as landowners that irrigate their field. This does not ensure, however, that such an authority will in any way be able to better represent the interests of the majority of the people, and in particular the poorest, than the government has done over the past few decades.

[87] ibid s 14.
[88] eg ibid s 12(6)(a).
[89] ibid s 11(d) following s 11(c) that focuses on the need for equitable distribution of water.
[90] This is evidenced by the criteria that members of the Authority must meet. See Maharashtra Water Resources Regulatory Authority Act 2005, s 4(1).
[91] eg SP Sathe, *Administrative Law* (New Delhi: Butterworths/Lexis-Nexis, 7th ed. 2004) 350.

Another problem arises from the naming of the MWRRA as 'regulatory'. Indeed, if it was truly regulatory it would have the power to set criteria for all water uses and focus as much on social aspects as environmental and economic aspects. In fact, the mandate of the MWRRA is focused on a relatively narrow set of issues. It is in no way the equivalent of an 'independent' water ministry or department that oversees all the various dimensions of water use. This is an issue because the adoption of the MWRRA Act gives the impression that it takes care of all water related institutional issues. In fact, the problem is that it gives prominence to the issues that are specifically addressed in the act and indirectly relegates other concerns, among them the realization of the human right, to a secondary plane.

This problem of terminology is accentuated by the fact that the establishment of the MWRRA arises in a context where large segments of existing water law are unclear and where the whole of water law is being rethought. Since the MWRRA becomes by default the focal point of hopes and contestation, it acquires a new sense of purpose that is unrelated to its real mandate and to what the Legislative Assembly may have wanted to do when adopting this act. Indeed, the realization of the human right to water remains an absolute priority because it is unfulfilled for many people. The utmost priority is thus for legal and institutional reforms that ensure the realization of this right, something that the MWRRA is neither tasked nor able to achieve.

3. Institutional reforms in perspective

Ongoing institutional reforms are bringing some of the most significant changes that the water sector has experienced in decades. They thus deserve particular scrutiny because of the extent of the transformation and the impact they will have for many years.

The setting up of water regulatory authorities is closely linked with the implementation of water sector reforms and the same principles are emphasized.[92] This implies, for instance, that regulatory authorities bring in a series of novel features to the water sector. This includes different elements, such as a focus on river basin planning, incentives for the participation of private sector actors, delinking access to water from control over land and shifting power away from the government.

Firstly, river basin planning has been called for on various occasions in view of the shortcomings of existing arrangements.[93] This is due to the fact that the River Boards Act has failed to deliver expected benefits. The Act has not been

[92] Parallels can be found with reforms undertaken in other countries, for instance, with regard to entitlements. For South Africa, eg JAM Döckel, 'The Possibility of Trade in Water Use Entitlements in South Africa under the National Water Act of 1998', in S Perret, S Farolfi & R Hassan (eds), *Water Governance for Sustainable Development* (London: Earthscan, 2006) 35, 37.

[93] eg National Water Policy 1987, s 3(3), National Commission for Integrated Water Resource Development Plan, Report (New Delhi: Ministry of Water Resources, 1999) 218 calling for the adoption of an Inter-State Rivers and River Valley (Integrated and Participatory) Management Act and National Water Policy 2002, s 4(2).

applied and boards have not been set up. Further, even if they had been set up, these boards would have had mostly advisory competences and failed to provide a framework for basin-wide planning.[94] This explains why one of the first recommendations of the World Bank in its 1998 report was to strengthen this act at the national level.[95] The setting up of regulatory authorities focusing on river basin management thus constitutes the answer in this area.

Secondly, the call for the participation of private sector actors in the management of water informs the setting up of water regulatory authorities in some direct and some more diffuse ways. For instance, while the MWRRA Act does not address the question of water privatization directly and, in fact, does not mention the term at all, it provides a broad framework that will facilitate some forms of privatization in the future. Additionally, the act must be read together with the State Water Policy to which it specifically refers.[96] The latter has a specific section on private sector participation which calls for the 'full and effective participation' of commercial enterprises and water service providers.[97] In other words, while the text voted by the Legislative Assembly is silent on privatization, it directly proceeds from another text which includes privatization as one of its specific aims.

The introduction of entitlements is one of the instruments in the MWRRA Act that will contribute to a profound reorganization of the water sector and will pave the way to certain forms of private sector involvement in water management. The crucial role of the entitlements scheme is acknowledged by the World Bank, which sees it as being at the heart of the MWRRA Act.[98] As per the definition given by the act, an entitlement is an authorization by any river basin agency to use water.[99] Additionally, the setting up of entitlements paves the way for trading in entitlements. This is also specifically provided in the act, which gives the MWRRA the power to fix the criteria for trading in water entitlements.[100] This provides the basis for individual or private sector trading in water entitlements.

At present, pilot projects are being implemented to demonstrate the viability of the new framework. One such pilot project is taking place in the Waghad irrigation project in Nasik district. In this area, high-earning cash crops such as sugarcane and grapes are grown. Additionally, a well-functioning system of water

[94] River Boards Act 1956, c III.

[95] World Bank, India—Water Resources Management Sector Review—Initiating and Sustaining Water Sector Reforms (Report No. 18356-IN, 1998) 57.

[96] Uttar Pradesh Water Management and Regulatory Commission Bill 2008, s 13(1) proposes to do the same.

[97] Maharashtra State Water Policy 2003, s 2(2)(5).

[98] World Bank Website, Empowering Users by Giving them Clear Water Entitlements, available at <http://go.worldbank.org/6TY7X9U3H0>.

[99] Maharashtra Water Resources Regulatory Authority Act 2005, s 2(1)(i).

[100] Maharashtra Water Resources Regulatory Authority, Procedure for Regulation & Enforcement of Entitlements 2007.

user associations had been in place for a number of years, the first ones being set up in 1991.[101] To this promising start, a generous package, part of the ongoing World Bank project that offers rehabilitation of the dam, main canal, branches, and minors has also been provided.[102] On the whole, the pilot project thus seems to be off to the best possible start and water users that benefit from the irrigation system in place seem generally pleased. Yet, while the allocation of bulk entitlements among water user associations in the command area is particularly welcomed by tail end users, this does not extend to the introduction of individual entitlements. Two main concerns are highlighted by local farmers. Firstly, individual trading could lead to transfers to the richer farmers rather than to the ones best able to put the water to effective use for the community as a whole. Secondly, unmonitored trading might allow a corporate entity to buy water entitlements in bulk, leading to a loss of control at the local level. In other words, even in a situation where all variables that can be controlled have been adjusted to ensure the success of the new scheme, local users are unlikely to willingly take up the idea of individual trading.

Thirdly, one of the main aims behind the adoption of the MWRRA Act is to sever the link between control over land and control over water. This is noteworthy because the nexus between land rights and access to water is socially inequitable and environmentally unsustainable. Yet, the way in which the break with traditional water law principles is proposed is likely to lead to conflicts with some of the fundamental principles of water law. In particular, the notion that water entitlements can be privately traded is difficult to reconcile with the recognition of water as a human right and a public trust. In the case of the public trust doctrine, the state may argue that as a trustee it has determined that privatization is the best way to sustainably manage water in the twenty-first century. This position would require the setting up of safeguards that are not in-built in an act like the MWRRA Act. In other words, while the setting of tradable entitlements may be justifiable under the public trust, it needs to be conceived in a broader context, which recognizes the hierarchically superior position of the human right to water. This reduces the margin of appreciation of the trustee in governing water.

Fourthly, the setting up of water regulatory authorities is premised on their capacity to take on some of the functions carried out by the state. Here, two potentially opposed trends can be identified. On the one hand, water sector reforms are premised on limiting the state's ability to control and provide in favour of a more limited role of facilitation. The promotion of decentralization,

[101] Also GB Keremane & J McKay, 'Self-Created Rules and Conflict Management Processes: The Case of Water Users' Associations on Waghad Canal in Maharashtra, India' (2006) 22/4 *International Journal of Water Resources Development* 543.

[102] Maharashtra Water Sector Improvement Project, Procurement Packaging Plan and Implementation Schedule for Civil Works of Participatory Rehabilitation & Modernization Component (Report No 41558, 2006).

community participation, and private sector participation are part of this broader agenda. On the other hand, the very changes that are proposed require new forms of control and regulation. This is due to the fact that no amount of water sector reforms can alter the basic nature of water services which makes water unsuited for forms of privatization that would leave the market to regulate itself. This reflects the natural monopoly nature of water which necessitates a monopoly even under conditions of privatization.[103] Thus, competition among private sector water service providers only occurs at the level of the allocation of a contract, not during its implementation, since there can be no more than one set of pipes in a given locality. In other words, there is competition for the market rather than competition in the market. This kind of privatization can only function to the satisfaction of water users with a strong regulator that has the power to ensure that the monopoly position is not abused. Additionally, even where water sector reforms seek to turn water into a commodity, this will never entirely alter its basic nature which is to be largely beyond human control since most of the water we use is part of the limited amount of water that circulates through the global water cycle. As a result, the setting up of regulatory authorities does not actually proceed from the logic of less bureaucratic control but of a different type of control.[104] There is, however, no guarantee that this new form of regulation will be either more effective or more equitable. This is illustrated in the case of legislation that specifically bars any appeal to civil courts, thus assuming that the new institution is better suited than the government and the courts to deliver socially equitable, environmentally sustainable, and economically efficient outcomes.[105]

C. Groundwater Legislation: An Old Model in a New Context

As indicated in chapter 2, the basic operating framework for groundwater is based on common law rules that give landowners dominant control over this water. Additionally, the tendency of policy makers to think of surface water and groundwater separately implies, for instance, that principles that apply to surface water do not necessarily extend to groundwater. This is, for instance, the case of the notion of public trust that was extended by the Supreme Court only to surface water.[106]

Groundwater regulation is the most pressing and challenging part of water law reforms at present. This is due to the fact that groundwater is now the main source of water for most water users and that the current outdated framework can

[103] eg J Budds & G McGranahan, 'Are the Debates on Water Privatization Missing the Point? Experiences from Africa, Asia and Latin America' (2003) 15/2 *Environment & Urbanization* 87.

[104] This is also the case in case of outright privatization as in the case of England and Wales. eg KJ Bakker, *An Uncooperative Commodity—Privatizing Water in England and Wales* (Oxford: Oxford University Press, 2003) 36.

[105] Uttar Pradesh Water Management and Regulatory Commission Act 2008, s 31.

[106] *MC Mehta v Kamal Nath* (1997) 1 SCC 388 (Supreme Court of India, 1997).

do little more than adjudicate claims that may arise between two landowners over their respective use of groundwater under their plot and in its vicinity. The challenge that groundwater poses has been recognized for quite some time, as witnessed by the fact that the Union government already put out a model bill for adoption by the states in 1970. This relatively early date of adoption of the model bill is reflected in its approach to groundwater regulation. Indeed, in the early 1970s, there was comparatively little discussion of the need for control by panchayats over natural resources or water and environmental concerns had only just made an appearance on the agenda of policy makers. It is thus not surprising to find that the 1970 model bill reflects the concerns and perceptions of that period. What is more surprising is that, despite several revisions, the model bill (re)proposed in 2005 is still based on the same premises.

Groundwater law reforms are noteworthy for several reasons. Firstly, the proposed changes conform to a model that is neither directly in line with water sector reforms nor influenced by the 73rd constitutional amendment, human rights, and environment principles. Secondly, they perpetuate the sectoral treatment of surface and groundwater in the twenty-first century, perpetuate a system that links access to groundwater and land and fail to acknowledge that groundwater is the primary source of drinking water and thus primordial in the realization of the human right to water. Thirdly, they constitute the only major law reform currently ongoing which is not directly influenced by water sector reform principles. This is positive insofar as it will ensure that different models of water governance persist in India but will at the same time create additional uncertainty because it perpetuates sectoralism.

1. The proposed reform model

A model bill for groundwater regulation was first proposed by the Union government for adoption by the states in 1970. It has been revised several times but the basic framework of the latest 2005 version retains the basic framework of the original bill. The reasons for the repeated attempts to foster the adoption of groundwater legislation by states can be explained by different factors.

Firstly, the political implications of groundwater use have become more significant with every passing year and with the increased importance of groundwater for irrigation and drinking water. State governments are thus wary of tackling the existing status quo. As a result, individual states refrained from introducing legislation for as long as they could to avoid generating confrontation with political vote banks.[107] Secondly, for similar reasons, the Union government has shied away from introducing legislation through the route used for the adoption of the Water Act 1974. Thirdly, it was technically possible for a while to simply dig further and further to access groundwater. Governments thus

[107] eg Narayanamoorthy & Deshpande (n 26 above) 37.

often found it easier to increase power subsidies than to regulate groundwater use. The limitations of such strategies have convinced an increasing number of states to legislate.

Recent legislative activity by states indicates that they are generally ready to follow the framework provided by the model bill. This is the case of states adopting a general groundwater legislation like Kerala,[108] or states focusing on its drinking water aspects like Karnataka, Madhya Pradesh, and Maharashtra.[109]

The basic scheme of the model bill is to provide for the establishment of a groundwater authority under the direct control of the government. The authority is given the right to notify areas where it is deemed necessary to regulate the use of groundwater. The final decision is taken by the respective state government.[110] There is no specific provision for public participation in this scheme. In any notified area, every user of groundwater must apply for a permit from the authority unless the user only proposes to use a handpump or a well from which water is drawn manually.[111] Wells need to be registered even in non-notified areas.[112] Decisions of the authority in granting or denying permits are based on a number of factors which include technical factors such as the availability of groundwater, the quantity and quality of water to be drawn and the spacing between groundwater structures. The authority is also mandated to take into account the purpose for which groundwater is to be drawn but the model bill does not prioritize domestic use of water over other uses.[113] Basic drinking water needs are indirectly considered since, even in notified areas, hand-operated devices do not require the obtention of a permit.[114]

The model bill provides for the grandfathering of existing uses by only requiring the registration of such uses.[115] This implies that in situations where there is already existing water scarcity, an act modelled after these provisions will not provide an effective basis for controlling existing overuse of groundwater and will, at most, provide a basis for ensuring that future use is more sustainable.

Overall, the model bill extends the control that the state has over the use of groundwater by imposing the registration of groundwater infrastructure and providing a basis for introducing permits for groundwater extraction in regions

[108] Kerala Ground Water (Control and Regulation) Act 2002.

[109] Karnataka Ground Water (Regulation for Protection of Sources of Drinking Water) Act 1999, Madhya Pradesh peya jal parirakshan adhiniyam 1986 and Maharashtra Ground Water Regulation (Drinking Water Purposes) Act 1993. On the Maharashtra Act, S Phansalkar & V Kher, 'A Decade of the Maharashtra Groundwater Legislation' (2006) 2/1 *L Environment & Development J* 67, available at <http://www.lead-journal.org/content/06067.pdf>.

[110] Model Bill to Regulate and Control the Development and Management of Ground Water 2005, s 5.

[111] ibid s 6.

[112] ibid s 8.

[113] ibid s 6(5)(a) only provides that the purpose has to be taken into account while Section 6(5)(h) which is the only sub-section referring to drinking water only considers it as an indirect factor.

[114] Model Bill (n 110 above) s 6(1).

[115] Model Bill (n 110 above) s 7.

where groundwater is over-exploited. It is the brainchild of an era that promoted governmental intervention without necessarily thinking through all the checks and balances that needed to be introduced alongside. As a result, the model bill is not adapted to the current challenges that need to be addressed.[116] It fails to include specific prioritization of uses, does not specifically address the question of domestic use, does not differentiate between small and big users, commercial and non-commercial uses and does not take into account the fact that non-landowners/occupiers are by and large excluded from the existing and proposed system which focuses on the rights of use of landowners. It is thus surprising that states are still drafting acts based on this outdated model. What is required is legislation that recognizes that water is a unitary resource, that drinking water is the first priority as well as a human right, and that panchayati raj institutions must have control over and use of groundwater.

2. Assessing reforms in the states

Most states have either adopted groundwater legislation in the past decade or are in the process of developing it. While most states are yet to adopt legislation, the need for one seems to be generally acknowledged. However, in an interesting twist, a state like Punjab that has 85 per cent of its land under cultivation is not contemplating the adoption of groundwater legislation because of the impact it would have on farmers.[117] Instead, Punjab is proposing to give incentives for crop diversification, to invest in artificial groundwater recharge, to meter electricity supply in critical areas, and to promote micro-irrigation.

The states that have adopted legislation that specifically focuses on groundwater include Goa, Himachal Pradesh, Kerala, Tamil Nadu, and West Bengal.[118] They differ in their coverage since some apply only to notified areas while others apply to all groundwater.[119] As noted above, Karnataka, Madhya Pradesh and Maharashtra have adopted limited groundwater legislation focusing on drinking water.[120] The only state that has consciously put groundwater in a broader framework is Andhra Pradesh where the groundwater legislation directly links surface and ground water in a general context of environmental conservation.[121] Apart from a conceptually broader framework for groundwater regulation and

[116] For additional comments, Ground Water Management and Ownership—Report of the Expert Group (New Delhi: Government of India, Planning Commission, 2007).

[117] ibid 29.

[118] Puducherry and Lakshadweep, two Union Territories have also adopted groundwater regulation instruments, respectively in 2002 and 2001.

[119] S Koonan, 'Groundwater—Legal Aspects of the Plachimada Dispute' in P Cullet, A Gowlland-Gualtieri, R Madhav & U Ramanathan (eds), *Water Law at the Crossroads—National and International Perspectives With Special Emphasis on India* (New Delhi: Cambridge University Press, 2009) 158.

[120] Maharashtra is in the process of adopting a broader groundwater act.

[121] Andhra Pradesh Water, Land and Trees Act, 2002.

specific consideration of drinking water issues, the Andhra legislation addresses groundwater in a similar manner to other groundwater acts.

The main institutional innovation proposed in the groundwater acts and the Andhra legislation is the setting up of a new authority or cell made of government civil servants and members nominated by the government because of their expertise. The balance between civil servants and other members varies. In Goa, the act simply authorizes the government to nominate members without specifying their origin.[122] In West Bengal, the majority are civil servants. In Kerala only four of the thirteen members of the Authority are civil servants while the rest is made of a combination of people with different expertise.[123]

The authority set up under the act is then tasked with different functions, such as notifying areas of special concern and granting permits to use groundwater in notified areas.[124] Among the acts that specifically focus on groundwater, the West Bengal legislation is the only one that gives the Authority a broader mandate that includes the development of a policy to conserve groundwater and organizing people's participation and involvement in the planning and use of groundwater.[125]

Following on the steps of the model bill, most acts fail to give drinking water clear priority of use even though most acts devote specific attention to the issue of drinking water.[126] The Himachal Pradesh legislation stands out insofar as it imposes on the Authority to give first priority to drinking water.[127] Additionally, some instruments specifically indicate that the use of groundwater as a public drinking water source is not affected by any control measures.[128]

An important aspect of most of these acts is to avoid altogether the thorniest question, which is the legal status of groundwater itself. Most acts avoid direct statements on this issue but the very fact of promoting the setting up of institutions controlled by the government that can regulate groundwater use in indirect and direct ways reflect a conception of water that sees it as being under the control of the government. The Himachal Pradesh legislation is rather forthcoming in this regard since it specifies that users of groundwater in notified areas must pay a royalty to the government for its extraction.[129] Additionally, the government is not even bound to use this royalty for groundwater-related activities, thus reflecting an understanding that groundwater is a resource

[122] Goa Ground Water Regulation Act 2002, s 3(2).

[123] Kerala Ground Water (Control and Regulation) Act 2002, s 3(3).

[124] eg Himachal Pradesh Ground Water (Regulation and Control of Development and Management) Act 2005, ss 5, 7.

[125] West Bengal Ground Water Resources (Management, Control And Regulation) Act 2005, s 6(2).

[126] eg Goa Ground Water Regulation Act 2002, s 23.

[127] Himachal Pradesh Ground Water (Regulation and Control of Development and Management) Act 2005, s 7(3).

[128] Goa Ground Water Regulation Act 2002, s 9. Also Karnataka Groundwater (Regulation and Control of Development and Management) Bill 2006, s 1(4).

[129] Himachal Pradesh Ground Water (Regulation and Control of Development and Management) Act 2005, s 12(1).

controlled by the government.[130] This can be understood as an extension of the full control given by several irrigation acts adopted in the twentieth century to the government over surface water. It is, however, surprising for at least two reasons. Firstly, there has been no debate on the status of groundwater and such a major change would warrant in-depth consideration. Secondly, in view of the traditional distinction between ground and surface water and in view of the adoption of acts that maintain this distinction, there is no reason for adopting principles that do not apply any more to surface waters following the Supreme Court's decision to recognize surface water as a public trust.

Besides strengthening the control that the government claims over groundwater, the various acts adopt a non-confrontational strategy in refusing to tackle existing overuse of groundwater. Thus, in the main, acts provide for the grandfathering of most existing uses. This amounts to refusing to tackle the real problem affecting groundwater. Indeed, as long as it is landowners that have most control over groundwater, there will be no scope for groundwater regulation that is socially equitable and environmentally sustainable. There is no incentive in the common law rules or in the acts that are being adopted for landowners to use the water responsibly and equitably. There is also no mechanism to ensure that groundwater is shared with non-landowners. Further, without a broader perspective, no single water user has any reason to recognize environmental needs ensuring that all ecosystem functions are met in the long term.

The limits of the old common law regime and new legislative efforts are well illustrated in the context of the dispute between the Perumatty Grama Panchayat in Kerala and the Coca Cola Company. The controversy erupted after the panchayat that first granted the exploitation licence decided not to renew it because of the lowering of the water table in neighbouring properties, as well as decreasing water quality to the extent that the local government primary health centre had concluded that the water was not potable.[131] The issue was brought to the courts and is now pending in the Supreme Court. The two decisions given by judges in Kerala gave two opposed views of groundwater regulation. On the one hand, the first judge found that even without groundwater regulation, the existing legal position was that groundwater is a public trust and that the state has a duty to protect it against excessive exploitation.[132] Additionally the judge made the link between the public trust and the right to life.[133] It was thus recognized that a system which leaves groundwater exploitation to the discretion of landowners can result in negative environmental consequences. The next decision took a completely different perspective and asserted the

[130] ibid s 12(2).

[131] CR Bijoy, 'Kerala's Plachimada Struggle—A Narrative on Water and Governance Rights' (2006) 42 *Economic & Political Weekly* 4332.

[132] *Perumatty Grama Panchayat v State of Kerala* 2004(1) KLT 731 (High Court of Kerala, 2003).

[133] ibid.

primacy of landowners' control over groundwater.[134] These two contradictory decisions illustrate the need for a framework that effectively ensures the sustainability of use of groundwater and the prioritization of drinking water over all other uses. Reliance on old common law principles is only able to justify individualized control but cannot in any way provide a broader framework of analysis. The inapplicability of the groundwater legislation to this dispute was noted by the judges. However, what is apparent is not the fact that the new legislation is not applicable but the fact that it would not have provided a framework for a more socially equitable and environmentally sustainable decision. The application of the act to future similar disputes may clarify matters in terms of institutional decision-making but it would likely lead to results fairly similar to the decision of the second judge. What is needed is a radically new perspective, something that the first judge perceptively understood. The Supreme Court now has the chance to provide a boost towards a new framework for groundwater regulation.

a) Power and groundwater

The link between access to groundwater and power has rapidly become prominent alongside the introduction of tubewells. At the most basic level, once access to water is mediated through a tubewell, the availability of power, and in particular electricity, becomes a direct determinant of access to water.

Where electricity is available, the next important variable is its price. Where the price is very low (or where power is free),[135] water is easily accessible but the easy availability of power may contribute to over-consumption of water and depletion of existing water sources. Where the price is too high, this may have equity consequences since the likely impact will be that rich people get more access to water than poor people because they can pay. The relation between groundwater and electricity is, however, much more complex than a straightforward cost issue.[136] On the one hand, by making groundwater extraction economically viable for farmers, states have contributed to farm incomes and to agricultural production. This explains in large part the unwillingness of most state governments to upset the delicate equilibrium as long as they can avoid it. On the other hand, the unreliable provision of electricity at uncertain times often leads to wasteful use of water. Cost is thus not the only issue that matters in terms of the development of measures that ensure equitable and environmentally sustainable use of groundwater.

[134] *Hindustan Coca-Cola Beverages v Perumatty Grama Panchayat* 2005(2) KLT 554 (High Court of Kerala, 2005) para 43.

[135] Narayanamoorthy & Deshpande (n 26 above) 223.

[136] cf T Shah, 'Groundwater Management and Ownership: Rejoinder' (2008) 48/17 *Economic & Political Weekly* 116, 119, arguing that '[m]ost of India's groundwater anomalies can be resolved in quick time simply by abolishing electricity subsidies and metering irrigation wells'.

The link between access to groundwater and electricity implies that the availability of electricity and its price directly impact access to drinking water. Thus, electricity policies should in principle be directly tied to drinking water policies to ensure that electricity tariffs do not impinge on the realization of the human right to water. Yet, in a policy context where electricity is increasingly being privatized, it becomes difficult to conceive of linking electricity with the realization of the human right to water. The existing legal framework does not provide the basis for making the necessary linkages. Indeed, the Electricity Act is framed is such a way that electricity tariffs cannot be based on social policies. Thus, Section 62 specifically prohibits any preference to any consumer of electricity thereby barring any possibility of making the use of electricity for pumping drinking water the ground for a special tariff.[137] The only possibility for drinking water to be taken into account is where a state government specifically decides to subsidize a class of consumers.[138] This is insufficient in a context where states are enjoined to avoid providing subsidies since the basic principle of the new electricity regulatory framework is to foster competition and efficiency rather than social policy objectives.

The low price of electricity is also seen as directly causing overexploitation of groundwater. The policy solution proposed is thus to increase electricity tariffs to control over-consumption.[139] The need to take measures to control groundwater exploitation must be done in such a way that access to drinking water does not become a casualty of such measures. Villages surveyed in the district of Aligarh in western Uttar Pradesh provided, for instance, an indication of the problems that can arise. Wells that had been the main or only source of domestic water supply up to about 20 years ago had all been abandoned for a variety of reasons including falling water tables. This had an economic cost for villagers forced to invest in their own handpumps and in a number of places huge health implications have arisen since in many cases salinity has become a major problem. In the case of the village of Nurpur in Pattal block of Aligarh district, fluorosis, which was already an existing problem has now become a health emergency, as there does not seem to be a single source of uncontaminated water in the village.

The importance of power, and electricity in particular, in the regulation of groundwater has largely been sidelined in existing regulatory frameworks. On the whole, states have often failed to tackle the electricity-groundwater conundrum because of the political costs involved. Additionally, the new regulatory framework for electricity has made it difficult even for willing states to adopt socially and environmentally conscious electricity policies linked to groundwater. The two factors have led to a situation where groundwater

[137] Electricity Act 2003, s 62.

[138] ibid s 65.

[139] eg Government of Rajasthan, Sector Policy for Rural Drinking Water and Sanitation (Draft, August 2005) s 3(14).

regulation remains largely blind to the nexus between access to groundwater and electricity.

D. Contributions and Limitations of Ongoing Water Law Reforms

This chapter has analysed some of the reforms currently unfolding in India. A number of lessons can be drawn from this limited but representative sample before turning, in chapter 5, to the specific question of drinking water.

Ongoing reforms are both momentous and limited in scope. On the one hand, one of the aims of the proposed changes is to radically transform what is perceived as an outdated and inefficient system. The shift to an integrated perspective on water management that takes river basins as its unit of reference and the sets of institutional changes that see the establishment of water regulatory authorities at the state level and decentralization of some functions to the local level are part of these changes. Reforms are meant to provide a clear break with the past and to comprehensively reorganize the water sector. In fact, these reforms will likely be much more significant than the first generation of water sector reforms that often focused on initiating changes through project-specific interventions. The new wave of water sector reforms whose first point of entry is sometimes legal changes, has the potential to introduce changes that will not be easily undone or sidelined as in the case of discrete projects because new water laws contribute directly or indirectly to changing the basic principles of water law.[140] The adoption of legislation leading to the setting up of independent water regulatory authorities is the clearest indication that water law reforms are meant to have important long-term impacts beyond the term of water sector reform projects. This process of institutional reform is likely to lead to a number of changes, which cannot necessarily be identified today because the powers given to an authority like the MWRRA are broad-ranging.

On the other hand, ongoing water law reforms do not comprehensively rethink water regulation in all its dimensions. They rather focus on a limited set of reforms. The guiding principles of water law reforms are to a large extent the principles of water sector reforms. This has two main implications. Firstly, proposed changes are based on the perceived need to conceive water as an economic good in all its dimensions and on the need to ensure its tradability. The focus on economic concerns precludes the development of a broader conceptual framework that does little more than pay lip service to environmental and social concerns. Secondly, the principles of water sector reforms are

[140] cf World Bank, Project Appraisal Document on a Proposed Loan to the Republic of India for the Maharashtra Water Sector Improvement Project (Report No. 31997-IN, 2005) 6 arguing that 'reforms need careful nurturing and support for a reasonably long time frame of about 12–18 years for them to take root and be sufficiently entrenched'.

not legal principles. Adopting these principles as the basis for water law reforms thus implies that there is no fresh thinking on water law principles but a relatively mechanical application of largely economic principles to law. A similar point applies to the setting up of independent water regulatory authorities. As noted earlier, they have broad-ranging mandates and the capacity to introduce significant changes in the water sector. At the same time, these independent authorities do not and cannot replace government departments. The main problem is that the new institutions are seen as operating in parallel or even in opposition to government institutions that are seen as inherently unable to change. Whatever the ultimate aim of the reforms may be, it is neither sensible nor practical to expect that the new institutions can replace government institutions. As a result, they should have been conceived as new institutions meant to work much more clearly in tandem with existing ones.

Similarly, reforms in the context of irrigation water are limited. Measures proposed focus on the specific question of the management of existing irrigation infrastructure rather than on the adoption of new irrigation acts. This is problematic and prejudicial because this leaves outdated irrigation acts largely untouched. In other words, new interventions superimpose new principles and measures on older frameworks without providing the structure for ensuring the smooth operation of the different legal instruments.[141] While the operation of existing irrigation acts has often been unsuccessful from the point of view of social equity, water conservation and cost of operations, it cannot be expected that the new acts will ensure that the irrigation sector will be reformed overall. In addition, as indicated above, the process of decentralization and participation that constitutes the rationale for water user association acts is itself flawed because decentralization is not conceived within the constitutional scheme of democratic decentralization and because participation focuses largely on participation of landholders in the management of existing infrastructure rather than on giving local people control over all aspects of irrigation at the local level.

The picture that emerges from ongoing reforms is one of a patchwork of measures and principles that are in large part superimposed on existing principles. This is problematic because this patchwork complements a water law framework which is itself in dire need of coordination and unity. The kinds of legislative interventions that are taking place can be partly explained by the fact that no government feels politically strong enough to suggest completely wiping out existing water laws. Yet, by doing so governments ensure that water law remains muddled. While politicians may appreciate the lack of clarity of water law as this allows them more flexibility in practice, this is problematic from a legal perspective. For instance, where the water user association legislation in Madhya Pradesh fails to

[141] Uttar Pradesh Water Management and Regulatory Commission Bill 2008, s 1(4) proposes changes that are much more direct since it specifically indicates that the bill prevails over anything contrary in the Northern India Canal and Drainage Act 1873.

specifically discuss the fate of irrigation panchayats under the Irrigation Act, there are theoretically different types of bodies that coexist in the same context.[142]

While water law reforms are in large part influenced by the principles of water sector reforms, groundwater constitutes an important exception since the proposed framework harks back to a much older proposal, which shares little in common with water sector reforms. Links with water sector reforms may nevertheless exist since increased government control could provide a stepping stone towards partly delinking control over land and water and thus opening the door to trading water entitlements. Even if groundwater legislation is not antinomic with the principles of water sector reforms, it does not immediately derive from the same principles. The inappropriateness of groundwater reforms to address current challenges indicates the problem identified with ongoing law reforms is not due specifically to the fact that they are based on water sector reform principles. Rather, the lesson from groundwater is that reforms that are neither grounded in the basic principles of water law such as the human right to water nor focusing on the social and environmental dimensions of water regulation are bound to be inappropriate.

Another development, which does not necessarily conform with the principles of water sector reforms concerns the importance of state intervention in the water sector. In the context of groundwater, the proposed framework leads in fact to more state intervention than before in controlling access to groundwater. More significantly, while water sector reforms strongly advocate the withdrawal of the state towards a facilitator's role and advocate demand-led strategies, proponents do not seem to see any contradiction in promoting parallel inter-basin transfers on a massive scale. Thus, the interlinking of rivers project is a supply-side solution which will require massive government intervention.[143] It will require government funding on a scale, which will likely dwarf any savings that the state will achieve in imposing cost recovery on irrigators and domestic water users through the introduction of demand-led water management. There are thus fundamental contradictions in the various reforms currently proposed.

Water law reforms that start from the existing legal framework to propose an alternative would first begin by assessing the existing laws and principles, determine whether they can be interpreted in novel ways that do not require amendment and examine whether the new measures proposed are compatible with existing basic principles of water law. Assuming that everything is found to be in need of change, the first step would be to adopt a list of new principles valid throughout the water sector, indicate the reason for the changes and examine their

[142] In Karnataka, water user associations have been introduced through amendments to the Karnataka Irrigation Act 1965 rather than through a separate water user association legislation. Karnataka Irrigation Act 1965 as amended in 2000 and 2002.

[143] cf J Bandyopadhyay & S Perveen, 'Interlinking of Rivers in India—Assessing the Justifications' (2004) 39 *Economic & Political Weekly* 5307.

validity in terms of constitutional and international law compatibility. Existing water law reforms fail to achieve this. The adoption of water policies is no substitute. Indeed, water policies cannot replace laws that are the constitutionally sanctioned instrument through which legislatures set out general principles and priorities. Additionally, laws are not supposed to be in compliance with policies because the legislature has full and entire liberty to set out a legislative agenda on water within the parameters set out by the Constitution. The idea that policies set out the framework and that laws are simply there to operationalize those principles is thus erroneous. Another shortcoming of the use of policies as a way to set out the general regulatory framework is that they are non-binding and the principles they contain are by definition not enforceable or justiciable. There is an added problem of wording, as in the already mentioned case of policies that give priority to drinking water while opening the door for a reordering of priorities at the government's discretion.

The limited scope of ongoing water law reforms is probably best exemplified by the lack of interest displayed in providing a comprehensive framework for a new socially equitable and environmentally sustainable water law in India or in each of its states. Indeed, the first thing which is missing either at the state or Union level is legislation that brings together all existing water law principles, weeds out outdated colonial principles and puts in place measures to ensure the realization of the constitutive tenets of water law confirmed by the Supreme Court that find no direct realization in existing legislation. Ongoing water law reforms have chosen a rather different path, which is not necessarily the most appropriate. This is why chapter 6 seeks to identify some bases for a different set of water law reforms.

5

Regulation of Domestic and Livelihood Water

Drinking water is the most fundamental component of any water law and policy framework. This is due to the direct link between drinking water and the human right to water. It deserves separate treatment in the context of water law reforms because it has not been addressed in the new laws examined in chapter 4. This is not due to the fact that other specific laws have been adopted for drinking water in recent years but to the fact that drinking water has not been a central priority in law reforms. Indeed, while all documents studied in chapter 4 reiterate the position that drinking water is the first charge of the water sector, the laws that have been and are being adopted are not specifically concerned with drinking water.

This does not mean that no reforms have been undertaken concerning drinking water. On the contrary, since at least the middle of the 1990s, sweeping reforms have been introduced in particular in rural areas. Yet, the reform of the policy context for the provision of drinking water in rural areas has never been discussed in the context of any legislation. There has thus been no parliamentary oversight of the process. In the case of urban areas, some attempts at privatizing water services in certain cities have been and are being pursued. This has been initiated in some big cities like Bangalore and Mumbai and is being mainstreamed to big and medium cities through urban reform projects.[1]

This chapter is concerned not only with drinking water strictly speaking but also with domestic water. This is the amount of water necessary for each individual to cook, bathe, and ensure a life free from waterborne diseases. The definition of drinking water usually includes water for cooking, food preparation, and personal hygiene and sometimes water for livestock as well.[2] This broader

[1] Two of the most important schemes are the Jawaharlal Nehru National Urban Renewal Mission (JNNURM) for big cities and the Urban Infrastructure Development Scheme for Small and Medium Towns (UIDSSMT).

[2] Indian Standard, Drinking Water—Specification (Second Revision of IS 10500) Doc. IS 10500:2004 (2005) 3, Himachal Pradesh Ground Water (Regulation and Control of Development and Management) Act 2005, s 2(d) and Madhya Pradesh peya jal parirakshan adhiniyam 1986, s 2(a).

definition does not include water used for livelihood functions such as growing subsistence crops. Yet, this is a dimension that needs to be added to the definition of drinking water if it is to encompass all the fundamental needs that are the basis for a decent human life in rural and urban areas.[3]

A. Policy Context for the Provision of Domestic Water

The availability of safe water sources has dramatically increased in the past few decades. Thus, it has been estimated that coverage in rural areas increased from 18 per cent in 1974 to 43 per cent in 1996 and 94 per cent in 2004.[4] Yet, while there has been a clear positive trend, the extent of the progress remains unclear as there exist different estimates of coverage.[5] This is partly due to the way figures are put forward. Thus a background document for the eleventh plan puts the figure at 95 per cent coverage but also acknowledges that nearly 20 per cent of habitations have slipped back to being only partially covered under the norms in force.[6] The problem of habitations slipping back is acknowledged as one of the biggest problems in the rural drinking water sector and can be due to a variety of reasons from the groundwater level going down to quality concerns or new habitations resulting in reduced per capita availability.[7] Additionally, physical access to water does not necessarily indicate whether the water is safe. The Planning Commission thus identifies 217,000 habitations suffering from water quality problems, with 118,088 habitations suffering from excess iron, 31,306 from excess fluoride, 23,495 from excess salinity, 13,958 from excess nitrate and 5,029 from excess arsenic.[8]

The figures mentioned here may in fact underestimate the problem of access to drinking water. A habitation survey undertaken in 2003 found that 16.9 per cent habitations were not covered and an additional 25.8 per cent partially

[3] Some aspects of a broader conception of the human right to water are considered in ch 6.C.

[4] M Black with R Talbot, *Water—A Matter of Life and Health* (New Delhi: Oxford University Press, 2005), AJ James, 'From Sector Reform to Swajaldhara—Scaling up in India' (2004) 23/2 *Waterlines* 11 and World Bank, Staff Appraisal Report—Uttar Pradesh Rural Water Supply and Environmental Sanitation Project (Report No. 15516-IN, 1996).

[5] KV Raju, K Das & S Manasi, 'Emerging Trends in Rural Water Supply: A Comparative Analysis of Karnataka and Gujarat' in KV Raju (ed), *Elixir of Life—The Socio-Ecological Governance of Drinking Water* (Bangalore: Books for Change, 2007) 1 noting that while the NCAER 1994 survey found that about half of the villages did not have any sources of potable drinking water, official assessments at that point were that 80% of villages were covered. World Bank, Country Strategy for India (2004) 40 also notes that direct observation suggests the data substantially overstates access to drinking water and sanitation.

[6] Planning Commission, Towards Faster and More Inclusive Growth: An Approach to the Eleventh Five Year Plan 2007–2012 (New Delhi: Government of India, 2006) 69.

[7] Agenda Note for Conference of State Ministers In-charge of Rural Water Supply and Sanitation, Sustainable Drinking Water Supply Sanitation for All 2012 (New Delhi: Rajiv Gandhi National Drinking Water Mission, 2007).

[8] Planning Commission—Government of India, *Eleventh Five Year Plan 2007–12—Volume II—Social Sector* (New Delhi: Oxford University Press, 2008) 164.

covered.[9] This could well be more representative of the reality than the higher estimates. Indeed, even in a relatively prosperous state like Maharashtra, partially covered and not covered habitations constitute 56 per cent of all habitations,[10] while in Rajasthan, only 32 per cent of habitations have been provided with safe drinking water.[11]

Additionally, the definition of a habitation is not that of a house, but a locality within a village where a cluster of around 20 families (or 100 people) reside.[12] This indicates that even when 100 per cent coverage of habitations is achieved, a number of more isolated houses may still remain uncovered. This has been recognized as a problem and under the eleventh plan, an effort is being made in the case of SC/ST habitations to cover them under the Accelerated Rural Water Supply Programme (ARWSP) even if there are less than 100 persons.

In urban areas, while the situation is in the aggregate better than in villages, supply often falls short of norms. Thus, in Gujarat apart from Vadodara and Surat that provide more than the stipulated 150 lpcd, municipal corporations like those of Rajkot, Bhavnagr and Jamnagar provide only about 75 lpcd.[13] This is compounded by highly unequal distribution within towns. Thus, in Rajkot while slums get 18 lpcd, the wealthiest segment of the population gets 83 lpcd. In Ahmedabad, while the worst-off households get as little as 5 lpcd, the best served get up to 500 lpcd.[14] Other studies confirm a general relationship between asset wealth and access to water.[15]

Access to water has rapidly changed over time. In rural areas, groundwater provides around 85 per cent of domestic water supply.[16] In urban areas, groundwater is the main source of water for about a quarter of households but most (70 per cent) depend on municipal water supply.[17] Yet a majority (about 61 per cent) do not have access to water within their dwelling and have to transport it from the main source. Even households that have access to tap water usually depend—54 per cent of them—on a tap which is not inside the house.[18]

[9] Department of Drinking Water Supply, National Habitation Survey 2003 (New Delhi: Government of India, 2003).

[10] Rural Water Supply and Sanitation, Eleventh Five Year Plan—Approach Paper (2006) 39.

[11] Government of Rajasthan, Sector Policy for Rural Drinking Water and Sanitation (Draft) 2005, s 2(2).

[12] Norms for Providing Potable Drinking Water in Rural Areas, available at <http://megphed.gov.in/knowledge/standards/guiderural.pdf>.

[13] I Hirway, 'Ensuring Drinking Water to All: A Study in Gujarat' in KV Raju (ed), *Elixir of Life—The Socio-Ecological Governance of Drinking Water* (Bangalore: Books for Change, 2007) 74.

[14] ibid 82.

[15] D McKenzie & I Ray, Household Water Delivery Options in Urban and Rural India (Paper prepared for the 5th Stanford Conference on Indian Economic Development, 3–5 June 2004) Table 3.

[16] Rajiv Gandhi National Drinking Water Mission—Department of Drinking Water Supply, Submission to the National Advisory Committee (2005).

[17] P Bajpai & L Bhandari, 'Ensuring Access to Water in Urban Households' (2001) 36 *Economic & Political Weekly* 3774.

[18] ibid 3775.

1. Evolution of the law and policy framework for the provision of domestic water

The provision of drinking water is of primary importance in the context of water regulation.[19] Yet, neither the Union nor the states have been particularly proactive in developing legal frameworks that elaborate on the right of individuals and groups to have access to water, the quantities involved and its quality. In fact, the provision of domestic water is, from a regulatory point of view, a patchwork of policy documents at different levels and rules and regulations often adopted in the context of specific legislation defining the rights and duties of specific municipal corporations.

Drinking water has been a policy priority from the early years of the independent republic as indicated by the launch of the National Water Supply and Sanitation Programme in 1954.[20] This quickly expanded and by the late 1960s a national drinking water supply programme was in place under the Central Public Health and Environmental Engineering Organization (CPHEEO) set up in 1953 under the Ministry of Health. The CPHEEO migrated to the Ministry of Urban Development in 1973 and has since then been a technical wing of the ministry dealing with urban water supply. It handled both urban and rural water supply until the National Drinking Water Mission (now Rajiv Gandhi National Drinking Water Mission) was set up in 1986 to take over rural water supply.

From the outset, drinking water was considered separately from water for agriculture or water for health. This is a legacy of the policies followed in European countries from the nineteenth century where the distinction was born out of the specific conditions operating there, in particular the prevalence of rainfed agriculture.[21] In India, a different scenario obtains but the distinction has largely been maintained to-date in policies, if not in practice. Similarly, the development of drinking water services was driven by the legacy of the connection made in European countries in the nineteenth century between unsafe water and epidemic diseases. This explains why authority over the drinking water programme was first given to the Ministry of Health.[22] In fact, a health rationale still informs the first attempt at drafting a legislation concerning drinking water.[23] As a result, a compartmentalized view of drinking water has emerged. This has

[19] *Delhi Water Supply and Sewage Disposal Undertaking v State of Haryana* (1996) 2 SCC 572 (Supreme Court of India, 1996).

[20] Earlier efforts did take place as indicated by the fact that Kolkata got its first filtered water supply in 1865 but they were limited in scope and coverage. TR Lee, *Residential Water Demand and Economic Development* (Toronto: University of Toronto Press, 1969) 5.

[21] Black with Talbot (n 4 above) 18.

[22] ibid 19.

[23] Department of Drinking Water Supply, Draft Guidelines for Preparation of Legislation for Framing Drinking Water Regulations (2007) 31 stating that the primary aim of framing a legislation on the regulation of drinking water is the protection of public health.

significant consequences, for instance, where water sources are used for drinking water and irrigation.

Different issues arise in the regulation of drinking water. Water quality is key to ensuring the delivery of water that is not harmful to health. Existing standards are in principle applicable to all water supply. These include, for instance, criteria for bacteriological examination whose result must yield no E. coli.[24] These standards are, however, not absolute since they are only compulsory in the case of piped supply of water.[25] Since there is a broad dichotomy between rural and urban areas with regard to piped supply, the implication is that standards applicable in rural areas differ more often than not from the ones applicable in cities. Additionally, existing standards are not generally binding. Rather, they constitute best practices and desirable aims for all agencies concerned with drinking water supply. Thus, while water quality standards are central to the supply of potable water, at this point the regulatory framework remains at best hazy. Standards exist and practitioners all seem to have a reference point in standards set by the Bureau of Indian Standards, the CPHEEO or the WHO. Yet, it would be difficult to hold actors accountable for the violation of a specific standard. This important gap has been recognized by policy-makers and a first attempt at introducing a binding framework for water quality is ongoing. The existing proposal is to make state water supply authorities responsible for ensuring that water intended for human consumption does not constitute a danger to public health and complies with Indian Standard 10500 highlighted above.[26]

Whereas quality standards are at least in general similar for all areas, drinking water policies largely differ between urban and rural areas. As a result, the framework for water supply in urban and rural water is examined separately in the next two sections.

2. Urban areas policy framework

Cities have generally been considered separately from rural areas. This is partly due to the fact that cities were until relatively recently the exception to the norm, which was life in rural areas. Cities have also always been centres of power from the local to the national level and the introduction of a different set of measures for the provision of drinking water in urban areas is thus relatively unremarkable in the policy context in which they developed. Yet, the separate—and more favourable—treatment of cities raises a number of questions. This has become even more important with the rapid urbanization and the increasing pressure that cities put on the environment, natural resources in general, and water in particular.

[24] Indian Standard—Drinking Water—Specification (First Revision) (New Delhi: Bureau of Indian Standards, IS 10500: 1991) s 3(2)(1).

[25] ibid 3(2)(2).

[26] Department of Drinking Water Supply, Draft Guidelines for Preparation of Legislation for Framing Drinking Water Regulations (2007) ss 45–6.

Water consumption in cities is generally much higher than in rural areas. Thus, a study of some major and medium size cities found that people consume an average of 91.6 litres per day, with a range among the cities studied going from 77 in Kanpur, Uttar Pradesh to 116 in Kolkata.[27] While this is a very high figure in comparison with the rest of India, this average is lower than the recommended 100 to 300 litres proposed by the WHO for optimal access.[28] With a 100 litre benchmark, the same study found that 65 per cent of households were water deficient.[29] Water deficiency in cities is correlated with socio-economic status and the level of education. The water consumption activities of households show that water is used for a variety of basic purposes. These include bathing with 28 per cent of daily use, followed by toilet use at 20 per cent, clothes washing for 18.6 per cent, cleaning utensils for 16.3 per cent and less than 10 per cent for drinking and cooking.[30]

As against these figures collected from surveys, the policy framework of the Government of India has recognized 70 lpcd as the absolute minimum level that needs to be provided even in times of drought and even in the poorest colonies.[31] This is divided into 10 litres for cooking and drinking, 30 litres for bathing and sanitation and 30 litres for washing utensils and clothes. Besides the minimum level of 70 lpcd, a gradation has been provided among different kinds of towns. Small towns with piped water supply but without a sewerage system are meant to receive 70 lpcd. Cities with a sewerage system should receive 135 lpcd while big metropolitan cities should receive 150 lpcd.[32] The CPHEEO Manual makes it clear that these figures only apply to households with individual piped connections. Where water is provided through public standposts, the norm is of 40 lpcd like in villages.[33] Further, different norms are proposed by different bodies. Thus, the Bureau of Indian Standards suggests that 200 lpcd should be provided in cities with full flushing systems while the National Commission for Integrated Water Resource Development Plan suggested that by 2050 cities should receive 220 lpcd and rural areas 150 lpcd.[34] Additionally, different states interpret these norms differently. In Karnataka, for instance, the water policy provides that towns should get 70 lpcd, municipal council areas should get 100 lpcd and the biggest

[27] A Shaban & RN Sharma, 'Water Consumption Patterns in Domestiç Households in Major Cities' (2007) 42/23 *Economic & Political Weekly* 2190, 2192.

[28] G Howard & J Bartram, Domestic Water Quantity, Service Level and Health (Geneva: World Health Organization, 2003) 22.

[29] Shaban & Sharma (n 27 above) 2193.

[30] ibid.

[31] Government of India, Report of the National Commission on Urbanisation—Volume II (1988) 294.

[32] National Commission for Integrated Water Resource Development Plan, Report (New Delhi: Ministry of Water Resources, 1999) 63.

[33] Central Public Health and Environmental Engineering Organisation, Manual on Water Supply and Treatment (New Delhi: Ministry of Urban Development, 1999) 11.

[34] Shaban & Sharma (n 27 above) 2191 and National Commission for Integrated Water Resource Development Plan, Report (New Delhi: Ministry of Water Resources, 1999) 64.

city corporations 135 lpcd while the draft Rajasthan policy suggests 120 lpcd for major cities and 100 lpcd for other cities.[35]

The delivery of water by the government in urban areas is premised on two different principles. On the one hand, urban dwellers benefiting from individual connections have been charged a fee for the service that is provided.[36] In most cases, the fee charged does not cover the overall cost of providing the water but this is changing fast in the context of ongoing reforms. On the other hand, free community taps and/or handpumps have over time been provided in a variety of places. In fact, some acts specifically provide that the relevant authority has the power to provide public standposts free of charge.[37]

Water supply in cities is in principle the responsibility of urban local bodies. These are regulated in a variety of ways. Certain major cities such as Kolkata have their own water supply and sanitation act. In other cases such as in Uttar Pradesh, water supply in cities is regulated in part by the Water Supply and Sewerage Act and in part by the specific regulations applying to the type of cities, usually categorized according to population size. There is thus no uniformity in the treatment of water supply in cities throughout the country.

The central obligation that is imposed on municipalities is to provide drinking water. For example, in Uttar Pradesh municipalities have a duty to provide 'sufficient supply of pure and wholesome water'.[38] In addition, they must also maintain public wells.[39] The main responsibility for water delivery in cities is in the hands of city-specific jal sansthans whose main task is to provide 'wholesome water' to city dwellers.[40] While the obligation to provide is central to the functions of municipalities with regard to drinking water, it is not necessarily absolute. Thus, in New Delhi, while the Council has a similar obligation to provide a sufficient supply of pure and wholesome water, this is qualified by the fact that it must only do what is practicable and at a reasonable cost.[41]

Water supply providers are funded at least in part through cost recovery from water users. In Uttar Pradesh, jal sansthans are funded through two different routes. Firstly, they charge a water tax on each property. The only exceptions to this rule are for properties which are not within a specified distance, often 200 metres, from the nearest standpost or other waterworks maintained by the

[35] Karnataka State Water Policy 2002, s 4 and Rajasthan State Water Policy (Draft) 2008, s 1(2)(3).

[36] L Mehta, 'Problems of Publicness and Access Rights: Perspectives from the Water Domain' in I Kaul et al. (eds), *Providing Global Public Goods—Managing Globalization* (Oxford: Oxford University Press, 2003) 556, 562.

[37] Calcutta Metropolitan Water Supply and Sanitation Authority Act 1966, s 45(2) and New Delhi Municipal Council Act 1994, s 154.

[38] Uttar Pradesh Municipalities Act 1916, s 7(j).

[39] ibid s 7 (jj).

[40] Uttar Pradesh Water Supply and Sewerage Act 1975, s 24.

[41] New Delhi Municipal Council Act 1994, ss 11 and 147.

jal sansthan.[42] Secondly, a water rate for actual consumption of water is also imposed. This is, for instance, Rs 100 per month in Aligarh.[43] The power to charge water users is associated with a number of duties. These include the obligation to maintain a system of water supply through pipes and to provide water at a prescribed pressure during prescribed hours.[44]

An important feature of a number of regulatory frameworks is the possibility for the authority in charge to disconnect private water supply. In Uttar Pradesh, jal sansthans have the power to cut off water where users fail to pay their bills. While there is no uniformity in the answer that officials give concerning the carrying out of disconnections, the act provides that the jal sansthan has the power to cut off power to anyone who fails to pay a bill within 15 days of receiving it and it appears to be enforced at least in some places.[45] This possibility takes additional significance in a context where households with an individual piped connection are prohibited from taking water away from the premises unless water supply is charged by meter, thereby preventing solidarity among neighbours.[46] The power to disconnect private water supply is a regular feature in other acts.[47]

In addition to state-specific or city-specific regulation, certain cities are governed by more than one legal regime concerning water supply. A situation which is covered by a Union act is that of cantonment areas. An update of an older colonial legislation, the new Cantonments Act 2006 still provides for the separate governance of cantonment areas by a cantonment board. Under the act, one of the duties of the board is to provide or arrange for the provision of water supply.[48] This includes a duty to ensure 'as far as possible' that the supply is continuous throughout the year and that the water is 'fit for human consumption'.[49] The duties of the board are also limited to the extent that it is not liable for any failure of supply arising from accidents or so-called unavoidable circumstances such as drought.[50]

The existence of different legal regimes governing different parts of a same city, which like in Delhi includes three distinct areas, is problematic in principle and in practice. Firstly, drinking water being a fundamental right, there is no justification for treating different urban residents of the same city differently. Secondly, in an immense city like Delhi, sub-dividing water supply responsibilities by areas could in principle be useful. The current scheme is, however, not based on a rational

[42] Uttar Pradesh Water Supply and Sewerage Act 1975, s 55.

[43] Personal communication with Mr Dwivedi, Mukhya Nagar Adhikari, Municipal Corporation, Aligarh.

[44] Uttar Pradesh Municipalities Act 1916, s 228.

[45] Uttar Pradesh Water Supply and Sewerage Act 1975, s 72.

[46] Uttar Pradesh Municipal Corporations Adhiniyam 1959, s 270(2).

[47] eg New Delhi Municipal Council Act 1994, s 169 and Mumbai Municipal Corporation Act, s 279.

[48] Cantonments Act 2006, s 186.

[49] ibid s 186(2).

[50] ibid s 194.

decision to improve water supply delivery for the poorest or the areas of the city suffering from much lower access to water. Thirdly, the existence of a separate framework for the New Delhi area which benefits from much higher per capita availability of water than any other area of the city can only reinforce existing inequalities in access to water. These inequalities in consumption are, for instance, highlighted by the fact the New Delhi Municipal Council area is supplied 462 lpcd while the neighbouring wealthy area of South Delhi has a supply of 148 lpcd.[51]

In recent years, there has been renewed interest in urban water policies. This is partly linked to the increasing strain under which existing systems have been put with rapid urbanization and partly due to donor agencies' interest. This has resulted not only in a number of projects at the city level but also in the redefinition of urban-specific water policies. Thus, in Karnataka, a new policy was adopted in 2002 whose objective is to ensure universal coverage of water and sanitation services.[52] While the latter is not new per se, it incorporates significant novel features. This includes a demand driven policy which seeks to put the burden of request on people and a policy to ensure coverage only for solvable people. The only qualification to the full cost recovery principle is that all citizens are to be provided with a minimum level of service.[53]

The minimum level of service is not defined in the policy, as in the case of many other municipal laws or regulations, and must be inferred from the guidelines of the CPHEEO. While this can be an appropriate functional system, it is problematic for at least two broad reasons. Firstly, the lack of reference to any specific standards leaves the door open to the Union government deciding at any point to amend existing standards without warning or consultation. In a context where private sector participation is being encouraged,[54] this is problematic since it does not provide effective guidance on the standards that drinking water providers must comply with. This is not a theoretical concern in the context of the progressive implementation of reforms under JNNURM and UIDSSMT. Indeed, some drastic changes are proposed. These include for instance the phasing out of handpumps even in cities like Agra and Mathura where respectively 70 and 90 per cent of the urban poor rely on groundwater for their drinking water needs.[55]

[51] Government of Delhi, Economic Survey of Delhi 2001–2002, 115.

[52] Urban Drinking Water and Sanitation Policy (2002), in Karnataka Urban Infrastructure Development and Finance Corporation, Establishment and Operationalisation of Karnataka State Urban Water Supply Council (Draft Report, October 2006) para 3.

[53] ibid para 6.

[54] eg Ministry of Urban Development and Poverty Alleviation, Urban Water and Sanitation Services Guidelines for Sector Reform and Successful Public-Private Partnerships (2004) ii.

[55] Agra, Checklist for the 'Urban Reforms Agenda' under JNNURM, 62, 65 available at <http://jnnurm.nic.in/nurmudweb/MoA/reform_agra.pdf> and Mathura, Checklist for the 'Urban Reforms Agenda' under JNNURM, 60, 64 available at <http://jnnurm.nic.in/nurmudweb/MoA/Mathura_Moa.pdf>.

3. Framework for rural water supply

The provision of drinking water in rural areas has been a priority for governments for several decades. This is due to the fact that at independence the drinking water situation was comparatively worse in villages than in towns and that an overwhelming percentage of the population lived in rural areas. In terms of coverage, there have been dramatic improvements over time even though there remain many issues to be solved from uncovered habitations to habitations that slip back to being uncovered or only partially covered and the fact that 'partially covered' habitations are for all practical purposes not covered since the minimum level for identification as partially covered is only 10 lpcd.

The provision of drinking water is primarily the responsibility of states. Yet, the Union government has played a very important role in fashioning the policies that states apply and provided significant funding to ensure access to water in rural areas. The most important body at the national level is the Rajiv Gandhi National Drinking Water Mission (RGNDWM) that functions within the Department of Drinking Water Supply established in 1999 under the Ministry of Rural Development. It has been the key institution with regard the development of policies and the administration of the rural drinking water sector. Among the schemes it implements, the ARWSP, which is funded by the Government of India and state governments plays a central role. The ARWSP was first introduced in 1972. Apart from an interruption during the 1970s, it has been a central component of the government's attempts to ensure full coverage of all habitations throughout the country. It continues to provide the basis for the Union government's interventions in rural drinking water.

The ARWSP Guidelines provide the core framework used by the RGNDWM in ensuring the provision of drinking water to all habitations in the country.[56] The Guidelines provide several key policy elements. Firstly, they define the different levels of coverage. Non-covered habitations are defined as having access to less than 10 lpcd. Partially covered habitations are those having access to 10 to 40 lpcd. Covered habitations are defined as having access to 40 lpcd. The figure of 40 lpcd used to determine the minimum level of coverage necessary to define a habitation as covered, has been determined through an amalgamation of figures for different basic minimum uses of water. These include 3 litres for drinking, 5 litres for cooking, 15 litres for bathing, 7 litres for washing utensils and the house and 10 litres for ablutions.

Quantity itself is not the only criterion to determine whether a habitation is covered. The source of water also needs to be within 1.6 km or 100 metre elevation in mountain areas. The water should also not be affected by quality

[56] Government of India, Accelerated Rural Water Supply Programme Guidelines (1999–2000) [ARWSP Guidelines].

problems even though no specific standards for determining quality are included and must thus be indirectly inferred from existing standards.[57] Another criterion is that a given public source of water such as a handpump should not be used to serve more than 250 people.[58]

The ARWSP Guidelines also acknowledge the direct link between drinking water for human beings and water for cattle. Consequently, in a certain number of states especially affected by drought, the guidelines mandate that an additional 30 lpcd should be provided for cattle.[59]

The minimum level of 40 lpcd is acknowledged by the RGNDWM as a minimum level of coverage which should be increased over time. Thus, in states where all habitations have been covered at the level of 40 lpcd, the Government of India has approved that the next level of service should be 55 lpcd within 500 metres of the house or 50 metre elevation in mountain areas.[60] Further, some states have long-term objectives which go beyond these minimums. Thus, Gujarat's Vision 2010 envisages, for instance, the supply of 80 lpcd in rural areas while the draft water policy in Rajasthan suggests 60 lpcd for rural areas in general and 70 lpcd for desert areas.[61] In Uttar Pradesh, the state proposes to provide 55 lpcd without house connections and 70 lpcd for households with house connections.[62] Thus, even within the rural sector there is no uniformity in the standards proposed.

B. Recent Reforms for the Provision of Domestic Water in Rural Areas

This section focuses on policy changes concerning the provision of drinking water in rural areas to illustrate the rapid and extensive changes that have taken place over the past few years. This focus is occasioned by two factors. Firstly, the majority of the population lives in rural areas and thus these require specific attention. Secondly, research interest has focused mostly on urban areas and the privatization of water services in cities. There is thus a need to provide more extensive research perspectives on the situation in rural areas.

The conceptual framework for drinking water policy has comprehensively changed over the past decade. A number of initiatives have been taken in different contexts from the Union level to international funding agencies' projects and state

[57] On water quality, see p 142 above.

[58] ARWSP Guidelines (n 56 above) ss 2(2)(3) & 2(3).

[59] ibid s 2(2)(2).

[60] Rajiv Gandhi National Drinking Water Mission—Department of Drinking Water Supply, Submission to the National Advisory Committee (2005).

[61] Raju, Das & Manasi (n 5 above) 28 and Rajasthan State Water Policy (Draft) 2008, s 1(2)(3).

[62] Memorandum of Understanding Between the State Government of Uttar Pradesh and the Department of Drinking Water Supply, Ministry of Rural Development, Government of India (2007) s 5(iv)(d).

level measures. While each can be analysed separately—and this section highlights two particular initiatives, the Swajal project and the Swajaldhara guidelines—the pattern which emerges is overwhelmingly consistent. In other words, while there are different problems in different parts of the country, while different actors have been involved in policy changes, the response given by policy makers at all levels is substantially the same. This implies that at the level of formal policy making there is a general consensus on the basic problems affecting drinking water in rural areas and the basic solutions that need to be adopted.

Changes in drinking water rural policies have been brought in a number of ways. These include changes in the existing policies of the Government of India, adoption of new policies at the Union and state level as well as development projects such as World Bank projects. This section first highlights some general changes in rural water policy and then specifically examines the case of the Swajal project and the Swajaldhara guidelines, two landmarks in the development of a new rural water policy framework which is still evolving.

At the Union level, one of the first important signs of the new conceptual framework is found in the 1999–2000 version of the Accelerated Rural Water Supply Programme Guidelines. They specifically highlight that one of the reasons for the existence of villages still uncovered includes the non-involvement of people in operation and maintenance.[63] The revision of the guidelines was specifically undertaken with a view to achieve full coverage of all rural habitations during the ninth plan (1997–2002). Three of the guiding principles are worth highlighting. These are the call for an increase in people's participation, the need to treat water as a socio-economic good and the use of 20 per cent of available funds for states promoting reforms along these lines. The revised guidelines make it clear that it is necessary to move away from the perception of water as a 'social right' and rather manage water as 'socio-economic good' to ensure its 'effective use'.[64]

The guidelines put significant emphasis on the need for people participation as a way to move away from supply-led to demand-led schemes. They identify a number of key conditions for the introduction of demand-led projects. These include ownership of assets and involvement in the setting up of the infrastructure. More significantly, the guidelines recognize that demand-led schemes require an imposition on people to pay for operation and maintenance and the knowledge that the government will not maintain the assets.[65] The message has recently been reinforced with the eleventh plan specifically calling for state support to panchayats for operation and maintenance as a 'hand-holding support for first few years before the local bodies become self-sustainable'.[66]

[63] ARWSP Guidelines (n 56 above) s 1(3).
[64] ibid s 3(1).
[65] ibid s 3(1).
[66] Planning Commission—Government of India, *Eleventh Five Year Plan 2007–12—Volume II—Social Sector* (New Delhi: Oxford University Press, 2008) 166.

The sector reforms put in place require all state and district authorities to impose at least 10 per cent capital cost payment by villagers.[67] Additionally, the ARWSP Guidelines lay down that the contribution of people must increase with the level of service provided. Thus, where villages want to increase their supply from 40 lpcd to 55 lpcd, they now have to pay at least 20 per cent of the capital cost on top of all operation and maintenance expenses.[68] The form of the contribution has been an ongoing debate. While in certain cases, a choice of cash, labour or materials is provided, some documents suggest a full cash contribution.[69]

The ARWSP Guidelines have significant influence on the policies followed in states. Yet, some states have gone further than others in adopting reforms in this field. Thus, Maharashtra has been particularly eager and was the first state to adopt demand-driven and participatory approaches for all its rural water supply service delivery already in 2000.[70] Besides, it also took the decision to phase out all government subsidies to local bodies for operation and maintenance of water supply.[71] The Government of Maharashtra has also gone much further than the ARWSP Guidelines in requiring not only 10 per cent capital cost contribution for a level of service of 40 lpcd (5 per cent for tribal settlements) but also 100 per cent contribution from the villagers for an increase to 55 lpcd, including for tribal settlements.[72] Maharashtra has also imposed that even the rehabilitation of existing drinking water supply schemes attracts a 10 per cent community contribution up to 40 lpcd and 100 per cent above that level. Other states have taken similar initiatives. Uttar Pradesh has, for instance, implemented the handing over of operation and maintenance of all drinking water schemes to gram panchayats. Rajasthan also has long-term plans to transfer ownership and responsibilities for the management of public water and sanitation assets to panchayati raj institutions as well as impose 100 per cent responsibility for operation and maintenance by users by 2012.[73]

Rural drinking water policy reforms have taken place at different levels. Initiatives taken at the Union or state level constitute two important elements of the overall reform process. Their effort must nevertheless be understood in the broader context of a string of water-related development projects funded, in particular, by the World Bank. Indeed, not only have World Bank projects been instrumental in pushing forward the new policy agenda but the World Bank has

[67] Planning Commission, Report of Working Group on Tenth Plan for Drinking Water Supply and Sanitation 2002–07, 4.

[68] ARWSP Guidelines (n 56 above) s 2(3)(1).

[69] eg Rajiv Gandhi National Drinking Water Mission—Department of Drinking Water Supply, Submission to the National Advisory Committee (2005) s 2(2). Note that below the poverty line, SC and ST families are excluded.

[70] World Bank, Project Appraisal Document—Maharashtra Rural Water Supply and Sanitation 'Jalswarajya' Project (Report No 26247-IN, 2003) 5.

[71] ibid 9.

[72] ibid 9.

[73] Government of Rajasthan, Sector Policy for Rural Drinking Water and Sanitation (Draft, August 2005).

also been closely associated with the policy changes taken at the state and Union levels. It advocated, for instance, already a decade ago that '[s]ubsidized water and highly centralized water management in the rural sector have resulted in poor water service at high cost' and that this undermined efforts to promote a more efficient and sustainable use of water.[74] It further advised the government that cost recovery was the only option to ensure that universal access to drinking water would not remain an 'unattainable dream'.[75] It also specifically called for the immediate imposition of operation and maintenance to users and the progressive implementation of capital cost recovery with the introduction of a 10 per cent contribution during what it saw as a transition period towards full cost recovery.[76]

Interventions by external agencies have evolved over the past decade. The project that broke away from the previous model of supply-driven, top-down drinking water delivery is the Swajal project that introduced the new framework represented in the AWRSP guidelines of demand-driven schemes, community participation, and communities bearing part of the capital costs and full operation and maintenance costs. Given its importance in informing the policy framework, it is discussed separately below. The Swajal project was not an isolated initiative. In fact, the principles proposed under this project are reflected in a series of subsequent projects. Projects such as the Kerala Rural Water Supply and Environmental Sanitation Project, the Second Karnataka Rural Water Supply and Environmental Sanitation Project and the Second Maharashtra Rural Water Supply and Environmental Sanitation Project have in particular attempted to remedy one of the shortcomings of the Swajal project in enshrining people participation in the context of the panchayati raj institutions. These projects also reflect the strengthened commitment to force villagers to pay for the capital cost of projects by introducing the principle that villagers must pay 100 per cent for any infrastructure that takes the supply beyond 40 lpcd.[77] This is in keeping with international policy guidelines that already called in 1992 for 'encouraging rural communities to undertake local development initiatives with the resources available to them'.[78] The most recent projects, such as the Uttaranchal Rural Water Supply and Sanitation Project, seek to further strengthen the role of the panchayati raj institutions in a move to further restrict the role of the Union and state governments in the actual provision of drinking water supply.[79] The

[74] World Bank, India—Water Resources Management Sector Review—Initiating and Sustaining Water Sector Reforms (Report No. 18356-IN, 1998) 25.

[75] World Bank, India—Water Resources Management Sector Review—Rural Water Supply and Sanitation Report (Report No. 18323, 1998) viii.

[76] ibid viii.

[77] Rajiv Gandhi National Drinking Water Mission—Department of Drinking Water Supply, Submission to the National Advisory Committee (2005) 6.

[78] International Conference on Water and the Environment, Report of the Conference (Geneva: World Meteorological Organization, 1992) 36.

[79] World Bank, Project Appraisal Document—Uttaranchal Rural Water Supply and Sanitation Project (Doc. IDA/R2006-0172/1, 2006) 17.

operating principle for the Bank is that of subsidiarity, implying that decisions should be taken at the lowest possible level.[80]

1. The Swajal project and related initiatives

Since the mid-1990s, a number of initiatives have been taken to foster better water supply and access to water in rural areas by following a new policy framework. It has by now been implemented in the context of several different schemes. While the instruments to deliver the new policy framework have evolved, the basic principles introduced in the mid-1990s continue to guide the overall policy framework for drinking water today.

One of the first formal steps towards introducing a new policy framework for drinking water in rural areas, was the implementation of the World Bank Uttar Pradesh Rural Water Supply and Environmental Project (Swajal project). This was a pilot project carried out in two regions of the then undivided Uttar Pradesh, Uttarakhand and Bundelkhand, both facing severe—but different—water supply problems. With a funding of $63 million, it covered about 1,200 villages in 19 districts between 1996 and 2002.[81]

The Swajal project was based around a string of important policy propositions. It sought to introduce a demand-driven approach to replace the supply-driven approach deemed to result in 'inefficient service delivery and poor quality of construction'.[82] The Swajal project thus sought to introduce participation by 'users' allowing them to determine their own contributions to the scheme and to manage operation and maintenance.[83] Under Swajal, participation by villagers had its own specific meaning. Indeed, it involved both ensuring people's control over schemes at the local level and introducing new obligations and responsibilities that villagers need to shoulder. Participation was also linked to the principle of cost recovery.[84] Thus, decentralization came in the form of people's control over some aspects of locally implemented schemes together with the imposition on villagers of the need to shoulder 10 per cent of the capital costs of new projects and the full costs of the operation and the maintenance of those schemes.

The ultimate rationale of the principle of cost recovery is that all projects should be entirely financially self-sufficient. At the outset, project proponents determined that the full cost recovery should only be imposed with regard to operation and maintenance. This was linked to the perception that that there would be sufficient 'demand' for this service while poverty might preclude

[80] ibid 17.

[81] The project size was reduced from an original $71 million after the 1999 mid-term review.

[82] World Bank, Staff Appraisal Report—Uttar Pradesh Rural Water Supply and Environmental Sanitation Project (Report No 15516-IN, 1996) 6.

[83] ibid 7.

[84] ibid 7–8.

'demand' for new expensive schemes in favour of maintaining or repairing existing infrastructure. Over the past decade the transfer of the responsibility of operation and maintenance to villagers has been mainstreamed and gram panchayats are now responsible for the operation and maintenance of the entire drinking water infrastructure.[85]

With regard to capital costs, a different strategy was adopted. In view of the political and practical impossibility to impose at once full cost recovery on villagers, it was decided to proceed incrementally. Thus, under the Swajal project a community contribution of 10 per cent of the capital costs was made mandatory. This contribution was divided into two components. The first was a 2 per cent cash contribution (1 per cent in the hill districts of Uttarakhand). The rest of the contribution could be in the form of labour, cash or a combination of the two. Anyone seeking an individual connection had to pay a cash contribution of Rs 1,000, half of which had to be paid upfront.[86]

This 10 per cent contribution was an arbitrary figure chosen as a way to introduce the principle of cost recovery rather than as a goal in itself. Thus, by 2002 a report prepared for the Planning Commission suggested that in all government projects 50 per cent of capital costs should be recovered.[87] Similarly a post-Swajal strategic plan of the government of Uttar Pradesh talks of progressively increasing cost sharing to 45 per cent.[88] This document is explicit in acknowledging that the figure is worked out according to its 'acceptability' and that reforms can only be introduced progressively.[89] Indeed a later document setting out a long-term vision for 2025 proposes to increase cost sharing to 50 per cent.[90] Further, the State Rural Drinking Water Policy already proposes to increase capital cost share to 20 per cent for coverage up to 55 lpcd.[91]

Another important facet of the Swajal project is that it sought to create new institutional capacity at the village level through the setting up of Village Water and Sanitation Committees (VWSC). These committees were first set up as subcommittees of the gram panchayat to which the panchayat had to delegate its powers to allow the committee to fulfil its responsibilities under the project.[92] The membership of the VWSC was in principle from seven to twelve members but the

[85] The World Bank, Implementation Completion Report for Rural Water Supply and Environmental Sanitation (Swajal) Project (Report No. 27288, 2003) 6.

[86] World Bank, Staff Appraisal Report—Uttar Pradesh Rural Water Supply and Environmental Sanitation Project (Report No 15516-IN, 1996) 142.

[87] GN Kathpalia & R Kapoor, Water Policy and Action Plan for India 2020: An Alternative (Delhi: Alternative Futures, 2002) 24.

[88] Rural Water Supply and Environmental Sanitation Sector Reforms—Strategic Plan (2002) 29.

[89] ibid 16.

[90] Long Term Vision 2025, in Memorandum of Understanding between the State Government of Uttar Pradesh and the Department of Drinking Water Supply, Ministry of Rural Development, Government of India (2007) annex.

[91] Uttar Pradesh State Rural Drinking Water Policy, s 3.

[92] Government of Uttar Pradesh, Order on Village Water and Sanitation Committee, No. 4430/33-1-95-373/95, 15 December 1995.

number was not fixed in the relevant government order. As a result, in the first phase of the Swajal project, while the VWSC were legally set up in the framework of the panchayati raj institutions, in practice they were independent from the gram panchayat. This was linked to the fact the World Bank did not think that panchayati raj institutions had the 'capacity and inclination to facilitate a demand-responsive approach'.[93] Rather, it was proposed that the project beneficiaries should be the ones directly involved in the schemes.[94] Efforts were made to ensure the representation of diverse constituencies by imposing, for instance, 20 per cent representation of SC/ST members and 30 per cent of women. However, on the whole, communities were to determine how to organize themselves.[95] The process of excluding the panchayati raj institutions and letting villagers organize themselves is problematic. Thus, allowing communities to determine how they want to organize themselves is more likely than not a way to reproduce existing patterns of power which tend to go along with caste and wealth. In a context marked by vast and even extreme inequalities, letting communities fend for themselves is likely to lead to further marginalizing of the weaker and poorer communities. Additionally, sidelining the panchayati raj institutions is also problematic because project beneficiaries are the people who have been able to pay the 10 per cent contribution at the outset. In practice, this largely restricts the number of project beneficiaries to the better-off members of the village. Nevertheless, it is the setting up of VWSC that was the visible face of participation in the first phase of the Swajal project.

The relationship between the VWSC and the panchayati raj institutions requires further comment. Indeed, while there was at first an attempt to link the VWSC to the panchayati raj institutions, in 1998 the Panchayat Act was amended and the VWSCs were then derecognized. This was partly linked to the perception that only committees made of elected panchayat members should fall under the panchayat and partly to the fact the project proponents thought the user committees would function better if they were completely independent from elected bodies. The severance of all links between the VWSC and the panchayati raj institutions quickly proved odd. Indeed, the constitutional mandate of the 73rd Amendment provides that panchayats control water supply at the local level.[96] In furtherance of this mandate, the Panchayat Raj Act was amended in 1994 and the functions of the gram panchayat now include the '[c]onstruction, repair and maintenance of public wells, tanks and ponds for supply of water for drinking, washing, bathing purposes and regulation of sources of water supply for

[93] World Bank, India: The Swajal Project, Uttar Pradesh (on file with the author, 2003) 2.

[94] World Bank, Staff Appraisal Report—Uttar Pradesh Rural Water Supply and Environmental Sanitation Project (Report No. 15516-IN, 1996) 15.

[95] Government of Uttar Pradesh—Uttar Pradesh Rural Water Supply and Environmental Sanitation Project (The Swajal Project), Description of Activities (DOA) of Support Organizations for the Batch 2 Planning Phase (on file with the author, undated) para 6(3).

[96] Constitution, art 243G and Eleventh Schedule.

drinking purposes'.[97] A water management committee has been provided to foster the realization of these functions. Since this did not leave any legitimacy for the committees constituted under the Swajal project, an additional layer of institutional complexity was added by allowing 'special invitees' to become de facto part of the relevant committee even though they are not formally elected. There can be up to seven invitees which brings the total number of people sitting in the committee to fourteen.[98] These invitees do not have the right to vote.[99] In other words, the non-elected user committees become indirectly part of the constitutionally sanctioned water committees of the panchayat. This is justified by the fact that 'users' are more directly involved in the relevant issues than the elected members of the panchayat committee. This is the same line of argument which is used to restrict membership in water user associations to landowners but lacks any legitimacy in the context of drinking water supply, which is without doubt the concern of each and every individual living in the concerned panchayat. A similar position has been adopted in the ARWSP Guidelines that request the setting up of VWSCs for each water project in reform mode made of members of the panchayat as well as co-opted members and other stakeholders.[100] The Guidelines also specify that this is a committee of the gram panchayat, regardless of its membership.

Swajal also initiated the progressive move towards reducing the government's contribution in drinking water schemes from being a 'provider' to being a 'facilitator'. Under Swajal, two initiatives were taken. Firstly, the Swajal project decided not to use existing government departments to implement the project but rather to use a Project Management Unit (PMU), an autonomous body established specifically to coordinate and monitor its implementation.[101] Secondly, the main interface between the PMU and the VWSC were so-called support organizations. The non-governmental organizations chosen to function as support organizations took on the tasks of facilitating specific schemes and working in individual villages. In practice, support organizations played a key role in the implementation of the Swajal project.

Official assessments of demand-led reforms were generally positive.[102] From the Union government's perspective, this can be assessed through two related

[97] Uttar Pradesh Panchayat Raj Act 1947, s 15(xi).

[98] Uttar Pradesh Panchayat Raj (Constitution of Committees of Gram Panchayats for Assistance in Performance of their Functions) Rules 2002, Uttar Pradesh Gazette, Extra., Part 4, Section (Kha) (13 September 2002) s 5.

[99] ibid s 5(2).

[100] ARWSP Guidelines (n 56 above) s 4(2)(3)(iii).

[101] Memorandum of Association and Rules—Uttar Pradesh Rural Water Supply and Environmental Sanitation Project Management Unit (PMU), in World Bank, Staff Appraisal Report—Uttar Pradesh Rural Water Supply and Environmental Sanitation Project (Report No. 15516-IN, 1996) 101.

[102] eg Ministry of Rural Development, All India Evaluation Study—Sector Reforms Projects in Rural Drinking Water Supply (2005) which provides recommendations on how to improve demand-driven rural water supply but does not call for a paradigm shift.

initiatives. Already in 1999 the Union government decided to broaden the Swajal experiment throughout the country. It started the Sector Reform Project (SRP) which sought to implement in 67 districts of the whole country the key principles of the Swajal project.[103] This was then extended to the whole country in the guise of the Swajaldhara Guidelines immediately after the completion of the Swajal project. Subsequently, SRP schemes were clubbed together with Swajaldhara projects.[104]

From the point of view of the World Bank, the Implementation Completion Report stated, for instance, that the establishment of the VWSC has allowed 'the village communities to fully participate in the process and gain a sense of ownership of the infrastructure schemes constructed under the Project'.[105] This success has led the Bank to implement several other projects based on the Swajal project philosophy in the past few years.[106]

Yet, even according to official assessments, a number of problems surfaced in the first phases of the reforms. Thus, in Uttar Pradesh, while the Swajal project was deemed successful, jal nidhi (the name given to the Sector Reform Project in Uttar Pradesh) proved unsuccessful. This has been explained as being due to the fact that in the plain districts where jal nidhi was implemented, demand for the schemes could not be created.

a) Swajal in practice—lessons from Bundelkhand

Evidence from Bundelkhand indicates that a more nuanced assessment of the results of the Swajal project is necessary.[107] It is first interesting to note the choice of the regions where the Swajal project was implemented. For different reasons, both Uttarakhand and Bundelkhand have faced many more difficulties with regard to access to drinking water than the alluvial plain areas of Uttar Pradesh. In a context where the government had generally failed to deliver, the demand for drinking water schemes was more acute in regions where water is less abundant. People could also more easily be convinced to pay for drinking water. The extent of deprivation—most acutely for dalits and deprived communities—can be judged from the fact that there are still villages where some people can be seen digging the dried riverbed mud to access drinking water as early as November. In

[103] World Bank, Implementation Completion Report—Uttar Pradesh and Uttaranchal Rural Water Supply and Environmental Sanitation (Swajal) Project (Report No. 27288, 2003) 4.

[104] Rajiv Gandhi National Drinking Water Mission—Department of Drinking Water Supply, Submission to the National Advisory Committee (2005) 21.

[105] World Bank, Implementation Completion Report—Uttar Pradesh and Uttaranchal Rural Water Supply and Environmental Sanitation (Swajal) Project (Report No. 27288, 2003) 4.

[106] eg World Bank, Project Appraisal Document, Maharashtra Rural Water Supply and Sanitation 'Jalswarajya' Project (Report No. 26247-IN, 2003).

[107] The following observations draw on meetings and visits in Chitrakoot and in the following villages of Chitrakoot district in the Bundelkhand area, Uttar Pradesh: Bandha, Barwara, Galshyapa, Jariha, Mayarahai, Murkatha, Patra and Piprodal.

Bundelkhand, unsuccessful government intervention has also led people not to expect the government to deliver.

This willingness on the part of at least some people to adopt the principles of Swajal cannot be confused with people's willingness to pay. In fact, a number of inter-related issues play out. Firstly, the poorest people who rely on water from common sources such as rivers, handpumps or wells are not directly influenced by cost recovery principles since they will only be affected the day their existing sources of drinking water are closed to them or fall into disrepair. In other words, as long as access to water per se is not affected, the poorest have nothing to gain and nothing to lose from demand-led schemes. They are unable to pay for them and do not benefit from them but are also not threatened and therefore have no reason to oppose them at the outset. In principle, their existing sources of drinking water will be maintained by the panchayat which now has the mandate to repair all existing drinking water infrastructure. In practice however, existing patterns of political and economic power at the local level may lead to emphasis being put preferentially on repairing the infrastructure that benefits mostly the rich rather than the handpumps and wells used mostly by the poor in the future. The poor would then be directly and dramatically affected by the new framework for delivering drinking water.

Secondly, there are situations where the poorest dalit communities paid and where the Swajal project was successful from the point of view of providing them access to water. The case of the hamlet of Charkauha in Unchadi village, Chitrakoot district illustrates this well. This hamlet of 17 dalit families is physically separated from the main village, has no electricity and is only connected to the main village by a dirt road. Their traditional source of drinking water was a well situated about 3 kilometres from the hamlet. This was for a time supplemented by a handpump that was eventually removed as the water stopped being potable. In Charkauha, people were ready to pay the capital costs imposed under the Swajal project, not because they thought it was right for them to pay for access to drinking water but because their desperate situation pushed them to welcome this opportunity. One of the reasons why villagers were convinced to pay and why the project was eventually successful is because the support organization that was involved in this case was Vanangana, a women's rights organization specifically working with dalit women in Chitrakoot district. This success story is thus not representative because most support organizations have little or no social commitment and are only implementing agencies.

Thirdly, willingness to pay is a relative concept related to one's ability to pay. Thus, the richer members of most villages are always happy to be given access to an individual connection and usually do not have reservations to the cost recovery principle. However, even for the wealthier members of villages, this 'willingness to pay' is always relative. While they could afford the upfront contribution they had to make and can pay the electricity charges that make up the bulk of monthly fees, nearly all of them agree that they would not be able to shoulder major repairs on their own. Indeed, in Bandha village of Chitrakoot district, the repeated failure of the pump had led the scheme to lay idle after hardly two years of use since the users were not able to pay the costs involved in the necessary repair. It has not functioned again. From this point of view, willingness to pay seems in most cases to be related to the way in which schemes are presented and to the extent of information provided to villagers as well as their actual understanding of the long-term consequences of the adoption of demand-led schemes.

Fourthly, the sustainability of the schemes is open to question. Where public funds are not available for repairs, the likelihood is that villagers will not be able to maintain the schemes when significant repairs are needed. In other words, community control is not sufficient to ensure the viability of the scheme in the long term because actual poverty prevents villagers from effectively maintaining the infrastructure. The village of Unchadi illustrates this point well. In this village, the scheme runs well and provides water to the users who have private connections. Yet, the electricity bill has gone up by about 30 per cent in the past couple of years in large part because the water table has gone down. Pumping the same quantity of water thus requires about 50 per cent more time. Rising electricity costs and the need to increase the depth of the borewell may signal the end of what is otherwise a success story because water users are not able to afford anything more than the current recurring charges.

Fifthly, there are concerns with regard to the public nature of the scheme. A first issue concerns public standposts. In villages where a piped system was put in place under the Swajal project, the schemes were usually implemented with a mix of private connections and public standposts. However, in almost all the villages surveyed, the public standposts had been closed a few years after the completion of the project. The reason given by people in charge of the project in the village was always that the public standposts had been closed because fees for the same had repeatedly not been paid. Another issue concerns handpumps installed on private land. In itself, this is not necessarily problematic. However, even if as in Mayarahai village such handpumps are generally accessible to all at present, this is still infrastructure that eventually becomes private property. Indeed, in some villages it was reported that after the end of the project, walls were erected around such infrastructure. Some people found it profitable to pay the whole 10 per cent of a given project to have it installed on their land. In this sense, the government gives a 90 per cent subsidy to a private individual with enough cash to pay the whole 10 per cent contribution. This is something which cannot easily be controlled when the government abdicates its responsibility to provide drinking water and leaves communities to battle it out among themselves.

Sixthly, the Swajal project does not seem to have contributed to reducing inequalities in access to water. In fact, the trend observed in the villages visited is that demand-led projects lead to increasing existing socio-economic inequalities in access to water. Richer people who are able to pay for the new infrastructure get improved access where they already benefited from better access than the poor.

The poor who cannot pay do not get any benefits from these new schemes. This is less so in the case of wells and handpumps but clearly so in the case of piped systems, as illustrated by the relatively fast closing of community taps. Indeed, richer people understand the principle very well. They clearly indicated that people who did not pay would not get access to the infrastructure. In other words, while Swajal is caste-blind, the fact that there are no mechanisms to ensure that the poorest and socially weakest benefit first from new schemes and from any funds available for operation and maintenance ensures that existing inequalities in access to water will most likely increase with time.

2. The Swajaldhara Guidelines

The Union government's positive assessment of the SRP and the Swajal project led to the formulation of the Swajaldhara Guidelines which extended during the tenth plan the key principles of the Swajal project to the whole country. 20 per cent of funds allocated to the ARWSP were directed to reform projects under the Swajaldhara Guidelines during this period.[108]

The Ministry of Rural Development spearheaded the introduction of Swajaldhara through the adoption of the Guidelines on Swajaldhara. The conceptual background is directly derived from the Swajal project. It first sets out to demonstrate that while water has been perceived as a social right, this is inappropriate as water should in fact be seen as a socio-economic good. Additionally, the delivery of the social right has been through the government which has not sufficiently taken into account the preferences of users and has been ineffective in ensuring the carrying out of operation and maintenance activities. This thus calls for a demand-led approach seeing water as an economic good.[109] The second paragraph of the background is even more revealing. It specifically links the transformation of a supply driven system to a demand driven system taking into account the preferences of users, 'where users get the service they want and are willing to pay for'.[110] This is taken one step further by indicating that it is the imposition of full cost recovery of operations and maintenance and replacement costs on the communities which are expected to generate a sense of ownership and ensure the financial viability and sustainability of the schemes.

The Swajaldhara principles are remarkably similar to the ones introduced under the Swajal project. Firstly, Swajaldhara provides for the adoption of a demand-led approach that includes participation of the community from the choice of the drinking water scheme up to its implementation. Secondly, the guidelines seek a form of decentralization and request that drinking water assets should be owned by the relevant panchayat and that the communities should have

[108] Ministry of Rural Development, Guidelines on Swajaldhara 2003, s 15(1).
[109] ibid s 1(1).
[110] ibid s 1(2).

the power to plan, implement, and operate all drinking water schemes. Thirdly, the participation and decentralization elements are brought together in the context of the financial principles which are a compromised version of full-cost recovery. Thus, while users have to bear the entire responsibility for the operation and maintenance of drinking water schemes, their contribution to capital costs is limited. In practice, this was first set at 10 per cent for a service level of 40 lpcd but,[111] in a number of situations, this percentage has already been exceeded. Under Swajaldhara, at least half of the 10 per cent contribution must be in cash, a significant increase over the 20 per cent under the Swajal project.[112] Exceptions have, for instance, been provided for scheduled tribes areas where the cash contribution was first reduced to one quarter of the community contribution.[113] Subsequently, in 2006 an amendment to the guidelines provided that the contribution in the case of villages where ST/SC constituted more than half of all habitations could be in any form without any stipulation of a contribution in cash.[114] Fourthly, from an institutional perspective, one of the consequences of a demand-led perspective is the rethinking of the role of the government. The guidelines here specifically provide that the aim is to shift the government's role from 'direct service delivery' to only supporting a limited number of activities such as planning, policy formulation, monitoring and evaluation.

An important aspect of the Swajaldhara scheme is that it was undertaken at the Union level without specific parliamentary mandate. Since water is largely a state prerogative, the states may not have bought into the new conceptual framework. The Union thus decided to proceed in two steps. Firstly, it decided to provide full funding for the scheme, a departure from the usual ARWSP norm where the Union and the states each share half of the costs. Secondly, it proposed that the states interested in taking up Swajaldhara funding should sign up a memorandum of understanding with the Union. The intent of the model memorandum of understanding circulated to states was to ensure that the reform principles would be, as far as possible, mainstreamed.[115] Apart from this general commitment to reforms of the drinking water sector, the states were, for instance, also called upon to hand over all existing drinking water schemes to gram panchayats for operation and maintenance.[116]

The process of decentralization and participation takes different forms under ongoing reforms. On the one hand, some of the proposed measures go towards

[111] ibid s 5(1).
[112] ibid s 5(3).
[113] Lok Sabha, Unstarred Question No 2451, Swajaldhara Yojana, Answered by Minister of State in the Ministry of Rural Development (Shri A. Narendra) on 18 March 2005.
[114] Government of India Department of Drinking Water Supply, Office Memorandum—Amendment to Swajaldhara Guidelines, Doc. No. W-11021/2/2003-TM.IV (SW), 15 May 2006.
[115] Draft Memorandum of Understanding between the State Government of __ and the Department of Drinking Water Supply, Ministry of Rural Development, Government of India (2003) s 5.
[116] ibid s 8.

ensuring that operation and maintenance of schemes is more successful. Thus, panchayats are for instance allowed to contract any required person where the government does not make its people available.[117] In principle, this should provide ways to ensure that any bottlenecks in the government do not affect actual operation and maintenance. Villagers complain, however, that only government officials have access to original spare parts and that private traders always provide sub-standard quality. On the other hand, the same set of reforms propose measures which are likely to lead to even more widespread inequalities in access to water within panchayats and particularly within different districts and states. Thus, the proposal seeking to allow panchayats to fix and collect water tariff is fraught with difficulties.[118] If tariffs are fixed at the level of each and every panchayat, the most likely consequence is that villages that suffer the most from water scarcity, for instance, because the water table is very low or water quality is low will have to bear all the costs themselves and will thus pay much more than villages that happen to be endowed with more or better water. Similarly, where costs are fixed within each panchayat without government overseeing, the likelihood is that dalits and other marginalized communities will be further marginalized because their say in the decision-making process will not only be low as it has been traditionally but more so because the reforms propose a decision-making process based on the notion of users which excludes most poor people. As noted above in the context of the Swajal project, where the communities are left to collect tariffs themselves, public taps are rapidly switched off. Whether the real reason is the one given by users that money is not being paid or whether it is a decision of a more political nature, the result is the same for people who are denied access to water. Given the nature of water, and drinking water in particular, any policy which does not attempt to redistribute the costs of getting access to water across social classes and across geographical areas is bound to fail from an equity perspective.

a) The Swajaldhara scheme in practice

The analysis of the Swajaldhara Guidelines gives an interesting overview of important issues raised by the introduction of the reforms in the drinking water sector. Yet, analysis of Swajaldhara on the ground is required because the successes and failures encountered in specific villages have important lessons for the development of legal and policy frameworks in years to come. The following analysis is informed by the situation in villages in Rajsamand and Bhilwara districts of Rajasthan, Badwani district of Madhya Pradesh and Chitrakoot district of Uttar Pradesh.

The first important finding is that the introduction of demand-led schemes is often welcomed by the richer and more powerful people in the village that

[117] ibid s 20.
[118] ibid s 8(vii).

understand the benefits they can derive from such schemes. In fact, such schemes may often replace richer people's dependence on their own private sources of water which in many cases are likely to cost more in the long run than a Swajaldhara scheme where the government subsidizes 90 per cent of the cost of the infrastructure. Thus, in Galshyapa village of Chitrakoot, the people who could afford to pay the Swajaldhara contributions were keen on this scheme because existing wells in their village were about 15 minutes away and the handpumps provided by the government were largely non-functional. Yet, this bright picture was undermined. Indeed, there did not seem to be any dalits among the users. As a result of this non-participation in the scheme, the dalits inhabitants of this village who have been relying on the local stream (in season) and on water obtained after digging the riverbed (throughout the period when the stream dries out) would carry on getting water in the same way as previously.

The demand-led approach is seen quite differently by poor people. The situation encountered in village after village provides the following general pattern. Whenever poor/marginalized people currently have access to 'a' source of water that satisfies their most essential needs, they are satisfied with what they have, however bad the situation may be, especially in summer months. Even where individuals think that better access through a new scheme would be beneficial, they often indicate that since they have not been able to pay the requested 10 per cent contribution, they are thus not part of the beneficiaries. Additionally, people often argue, especially in Rajasthan, that they believe water should be provided free by the government because it is such a basic necessity of life.

Swajaldhara projects thus raise questions of equity within villages since the professed willingness to pay that provides the conceptual justification for demand-led schemes is in fact a function of wealth. As a result, the poorest people who most urgently need better access to drinking water are the first to be excluded from the list of beneficiaries. In reform speak, the poor exclude themselves by not paying the 10 per cent contribution but this is a fallacy which is debunked by the reality on the ground. If anything, the ground reality indicates that people pay either when they can clearly afford it or when they are desperate enough to allocate more resources to water, necessarily at the expense of some other basic vital need such as food or health.

The issue of equity in access to water is not a new problem in Indian villages. Indeed, caste equations have, for instance, played a major role in differential access to water for centuries. While in formal legal terms, caste-related inequalities have been banned, in practice they still affect life in many places. Swajaldhara and the demand-led reforms do not affect caste equations directly. However, the problem is that they do not take into account the fact that caste discrimination is still prevalent in direct and indirect ways. As a result, demand-led reforms have the potential to introduce wealth-based inequality in access to water which may reinforce existing inequalities. Two examples of these trends can be given. Firstly, in the village of Bagatpura, Rajsamand district, the dominant Rajput people forced the dalit households to pay through threats even though the latter were economically too weak to afford the contribution. This went to the extent of a handpump being broken by upper caste people to force dalits living there to pay up the 10 per cent contribution. Secondly, in both Maharashtra and Uttar Pradesh, where villages are big and made of different clusters, the Swajaldhara scheme may never reach the dalit hamlet. This may or may not be a direct consequence of caste equations but the fact that Swajaldhara schemes may lead to such results as a result of a combination of long-standing caste-related issues together with the economically weaker position of dalits raises questions concerning the rationale of demand-led reforms.

The issue of intra-village equity is deeply rooted in the philosophy of the reforms which want to leave villages to manage their own water schemes. Given the prevailing condition in many parts of the country, it is impossible to expect that the wealthy, powerful, largely high caste members of the village will use the new concept of village ownership as a way to foster less unequal social relations within the village. Indeed, local politics may in fact dictate the contrary. Thus, in Sirola-Pithoda, Rajsamand district, people willing to pay the initial contribution were kept out of the scheme because they did not enjoy the favour of the powerful teacher who had initiated the scheme. Similarly, in Jogela-Miyala, Rajsamand district, a family that was willing to pay and whose house was close to the main supply pipe had been denied a connection because they did not enjoy the favour of the *sarpanch* who was also the committee president. This is particularly problematic when drinking water is the subject matter. Thus, the government's decision to progressively withdraw is likely to create, at least in the medium- and long-term, increased inequalities in access to water.

With regard to the selection of specific schemes, the principle underlying demand-led schemes is that users choose the scheme that best suits their needs and finances. In what are by any measure very poor areas of the country such as Rajsamand and Bhilwara districts, one would thus expect the gram sabha to decide in favour of the scheme that is the most cost effective and delivers benefits to the greatest number of people in the village. It is thus surprising to find that many villages opt for relatively expensive piped water schemes relying on a borewell (and its attendant electricity consumption) and a water tank to deliver benefits to private individual connections and community standposts. Indeed, according to Ministry figures, while the average contribution for an individual house connection is Rs 965 and Rs 810 for a community tap, it is only of Rs 412 for a handpump.[119] Despite this, the Ministry recorded that there were only 3,186 beneficiaries of handpumps and 5,451 for individual piped water connections.[120] While there are a number of places in the country where the water

[119] Ministry of Rural Development, All India Evaluation Study—Sector Reforms Projects in Rural Drinking Water Supply (2005).

[120] ibid.

table is either falling fast or has already fallen so low that handpumps may not be a technically viable solution, the relative lack of popularity of handpumps under Swajaldhara cannot be explained by technical factors. Indeed, in many parts of the country, including a very dry state like Rajasthan, it is the millions of handpumps installed by the government over the past few decades that actually provide access to water for hundreds of millions of people on a daily basis. Their relevance has never been put in doubt. While schemes are effectively chosen by the users, they often rely on advice given by outsiders, such as NGOs working as support organizations, since they may not have the necessary technical expertise to evaluate all different options. It is possible that advice provided to communities emphasizes the benefits of individual house connections. Whether that is the case or not, the choice of schemes is clearly influenced by VWSC membership. Indeed, field observation reveals in village after village that while the gram sabha is technically the body making the choice, in practice, the committee generally seems to be led by some of the more powerful and wealthier men in the village. This easily explains why so many villages would favour relatively expensive schemes because the decision is in effect taken by those 'users' and not by the actual meeting of the whole village. Another surprising observation is that despite the fact that one of the 'fundamental reform principles' of Swajaldhara is to promote conservation measures such as rainwater harvesting, none of the villages surveyed had either considered or implemented such structure.[121]

In institutional terms, the Swajaldhara guidelines tried to remedy the rapidly apparent democratic deficit arising from the setting up of VWSC besides the panchayati raj institutions. As a result, all VWSC were set up under the gram panchayat which, in deeply divided communities, is a precondition for any attempt to foster social and gender equity in the schemes taken up. Yet, the basic philosophy of the panchayati raj constitutional scheme is not fully upheld. Indeed, as noted earlier, in Uttar Pradesh, all VWSCs are in practice comprised of seven members of the gram panchayat water committee and seven members co-opted from among the users. These 14 people in effect constitute the committee that oversees drinking water projects, thus allowing users a controlling say in the decisions of the committee. This is confirmed by the fact that the ownership of public water schemes rests with the VWSCs.[122]

The panchayati raj institutions and their democratic nature also comes under attack from a different angle with the original Swajaldhara scheme. Assuming that the gram panchayat wanted to use its own funds to foster access to water, in the case of Uttar Pradesh, it was in fact prohibited from doing so. Indeed, panchayats and the state government were specifically prohibited from using any government

[121] Ministry of Rural Development, Guidelines on Swajaldhara 2003, s 3(1)(vi) and P Sampat, '"Swa"-jal-dhara or "Pay"-jal-dhara—Sector Reform and the Right to Drinking Water in Rajasthan and Maharashtra' (2007) 3/2 *L Environment & Development J* 101, 121, available at <http://www.lead-journal.org/content/07101.pdf>.

[122] Uttar Pradesh State Rural Drinking Water Policy, s 5(1).

fund as a substitute for people's contribution.[123] The intent of Swajaldhara to force individuals to pay whether they can afford it or not is also clearly stated in the guidelines' specific admonition that elected MPs and MLAs are prohibited from using the funds at their disposal for development work in their constituency to pay the community contribution.[124]

The importance attached to cost recovery under Swajaldhara implies that the costs borne by users are an important element to be taken into account in analysing the scheme from a policy perspective. Two main elements can be distinguished, the contribution to capital costs and the contribution to maintenance and operation costs. The limited contribution to capital costs was found in the Ministry report quoted above not to exceed on average Rs 985 for a private connection. In the villages surveyed, the actual figures were much higher. They ranged from a low of Rs 600 in Kansiya, Bhilwara to a high of Rs 7,000 Jogela-Miyala, Rajsamand.[125] This reflects the fact that users have to bear the cost of bringing pipes from the main line to their house and where the project design does not take the main line towards their house, their only option is to pay the full cost of their own connection from the main line. Another disturbing feature observed in a number of villages was that the cost of becoming a 'user' of the scheme shoots up for people who have not joined at the outset. The increase could be anything from 60 per cent to nearly 130 per cent.[126] This was usually explained by committee members as being an 'interest' that newcomers should pay on the original sum. This is one of the ways in which leaving communities to arrange their own affairs leads to increasing inequalities in favour of a group of people who are most often already the wealthier and politically astute individuals in the village.

The contribution to operation and maintenance raises a different set of issues. With regard to actual amounts paid, there are again significant discrepancies. The report of the Ministry finds an average Rs 37 paid for individual in-house connections per month.[127] Village surveys indicate wide variations from Rs 36 to Rs 150. Two important findings emerge from these figures. Firstly, variation seems to be explained mostly by the electricity cost associated with running the pump. The lack of transparency within villages makes it difficult to assess the causes of such differences. They may be ascribed partly to the depth of the water table but are also likely to be influenced by the actual amount of water people take. Indeed, while people generally seem to be reluctant to share water obtained through Swajaldhara schemes with their neighbours and thus limit consumption

[123] Ministry of Rural Development, Annual Report 2002–03, 130.

[124] Ministry of Rural Development, Guidelines on Swajaldhara 2003, s 5(4). See p 169 for some of the changes proposed in the meantime.

[125] Sampat (n 121 above) 117.

[126] ibid 117.

[127] Ministry of Rural Development, All India Evaluation Study—Sector Reforms Projects in Rural Drinking Water Supply (2005) vii.

to their individual household needs, some people freely acknowledged using their Swajaldhara connection to water their gardens and in at least one village the individual in charge of the scheme had ensured that he could irrigate his own fields as an indirect benefit of the scheme. Secondly, the costs paid by people for operation and maintenance never reflected the long-term depreciation of the scheme. In other words, the monthly cost paid by individuals seems to generally reflect the immediate electricity cost. This has immense implications where anything more than minor repairs need to be undertaken. Indeed, when asked the question directly, people are often quick to say that they would not be able to pay anything above the current fees. This is problematic because it indicates that the way the schemes actually function do not ensure their long-term viability. The emphasis is on ensuring that the schemes start to function, not on ensuring their long-term functioning. In the villages visited which have been plagued by various types of access to water problems, this was nothing extraordinary. Queries concerning access to water in case the scheme stopped functioning because the village could not afford repair costs usually have people simply indicating that in such a situation they would go back to their previous sources of water. Besides the kind of pessimism that may be apparent in such responses, a bigger problem surfaces. Where the main sources of drinking water before Swajaldhara are handpumps or other infrastructure built by the government, 'going back' to these sources seems an easy option today. In the future, when all operation and maintenance costs are imposed on villagers, such complacency will not be possible for anyone. In other words, when full cost recovery principles are implemented, villagers will be left with the choice of paying, whether they can afford or not, or not get access to water at all.

As indicated above, the introduction of the reforms is largely premised on the implementation of the constitutional mandate devolving powers to panchayats. While this is laudable in theory, practice indicates that a simple withdrawal of the government in favour of local actors is not progressive unless accountability at the local level is ensured and enforced by higher authorities. Where this is not the case, 'decentralization' is another word for increased concentration of power in the politically savvy and wealthier members of the local community. Two specific issues have come up in this regard in the surveys of Swajaldhara villages. Firstly, in no village were accounts available. The usual excuse was that they had been sent to auditors. Secondly, in most villages people paid money but rarely got receipts. Even where receipts were obtained, they were never provided information as to the use of this money.

The lack of external supervision is what allows the subversion of different aspects of the scheme by enterprising people. Thus, in some areas, contractors discovered in Swajaldhara a good business opportunity. They can pay the 10 per cent capital cost contribution on behalf of the community and in return get the government to pay 90 per cent of their costs. Subsequently, they can run the scheme according to their own preferences. On paper, the community has paid and the scheme is operated at the local level. This satisfies the basic principles of the reforms. In other villages, different ways to work around the scheme's principles were found. In Ker Kheda, Bijoliyan block, Bhilwara district, Hira Lal, *sarpanch* and chairman of the Swajaldhara scheme had paid the whole community contribution himself and had probably been paying the electricity bill as well. Two likely reasons explained this generosity. Firstly, the whole village was dependent on his borewell—the only one in the village—when other sources of water dried in the summer. The new scheme thus likely reduces his own direct costs of providing water in summer months. Secondly, water was identified by villagers interviewed as a major concern and everyone seemed extremely pleased with the new sources of water provided through this new scheme. This was likely going to strengthen his position politically in the village. This is not particularly surprising since similar outcomes were already reported a decade ago in early attempts to impose cost recovery on villages.[128]

A last feature is the issue of disconnections. In principle, there is no link between demand-led schemes and disconnections. Yet two issues arise. In a situation where operation and maintenance is a monthly feature, the payment of the electricity bill is a precondition for actual access to water. Even where the amounts appear small, the users may not be able to pay. Thus, in Mukund Puriya, Bhilwara, despite making use of the pump only 30 minutes in the morning and 30 minutes in the evening, individual users' share was about Rs 100 a month. Since most people could not pay this amount, the village ended up within about a year and a half after commissioning the scheme with an unpaid electricity bill of Rs 12,000 which led to their connection being disconnected. This raises important questions with regard to disconnection of water supply. In a context where disconnections have been confirmed as unacceptable even after full privatization in the UK, there is little doubt that disconnection from water supply is a violation of the human right to water.[129] In the case of Indian villages, the additional difficulty is that it is not water per se which is disconnected but the means of accessing it. Yet, in a country where an ever increasing share of drinking water comes from groundwater and where the water table falls so fast in so many places that tubewells are a necessity in an increasing number of villages, disconnecting electricity amounts to stopping people from having access to water. It is thus imperative to look at water supply in a broader light and do exactly the contrary of what some of the reforms urge us to do. Instead of disintegrating systems in separate units to make them all individually cost effective, what is in fact required is to consider systems in their broader context. Separating electricity consumption from access to drinking water is thus a complete misnomer in an increasing number of villages in India.

[128] A Behar, 'Revitalising Panchayati Rajs—Role of NGOs' (1998) 33/16 *Economic & Political Weekly* 881, 882.

[129] On disconnections and the human right to water, ch 6.C.2.

3. Swajaldhara in the eleventh plan

The implementation of the Swajaldhara guidelines in its first few years of operation led the government to rethink the scheme for the eleventh plan. This led to a series of documents proposing different ways to take the reforms forward.

In the first place, it was suggested that the Swajaldhara scheme would be discontinued at the end of 2007. This decision did not imply that the principles underlying the reforms were being abandoned. In fact, it had been proposed since at least 2005 to extend the reform principles of Swajaldhara to ARWSP from 2007 onwards.[130] In other words, the idea was to progressively mainstream the reform principles beyond the Swajaldhara projects to all rural drinking water projects. At the same time the break with Swajaldhara signalled a desire to rethink parts of the reforms in view of some perceived failures and resistance to the reforms. Indeed, official assessments of Swajaldhara found that it had been plagued by 'constraints' because government officials were slow to adopt reforms and because panchayats lack finances and skills to take up the responsibility immediately.[131] The government also conceded that difficulties in collecting the community contribution were a factor impeding the success of Swajaldhara.[132]

The combination of these two different assessments led to suggesting a revamped reform effort that sought to address some of the shortcomings identified by the Government in the implementation of demand-led reforms. Firstly, while the principles of Swajaldhara were to be upheld, the financing of projects was to change to the pattern implemented under ARWSP where states and the Union each contribute 50 per cent of the funding.[133] Further, the states were to be given the discretion to determine how and whether to foster community contribution. The incentives given to the states were that the community contribution would be deducted from the share of the state. Another relaxation concerned the community contribution. In villages where more than half the population is SC/ST, the stipulation of a cash contribution was to be abandoned.[134] An indirect way to relax the element of community contribution is to allow, for instance, MPs and MLAs to use development funds at their disposal for this purpose.[135]

[130] Rajiv Gandhi National Drinking Water Mission—Department of Drinking Water Supply, Submission to the National Advisory Committee (2005) 26.

[131] Rural Water Supply and Sanitation, Eleventh Five Year Plan—Approach Paper (2006) 3.

[132] Lok Sabha, Unstarred Question No 1550—Slow Pace of Swajaldhara Yojana, Answer of Minister of State in the Ministry of Rural Development (Shri Chandra Sekhar Sahu), 9 March 2007.

[133] Department of Drinking Water Supply, DO No W-11012/15/2007/DWS-III, 30 August 2007.

[134] Office Memorandum—Amendment to Swajaldhara Guidelines, No W-11021/2/2003-TM. IV (SW), 15 May 2006.

[135] eg Uttar Pradesh State Rural Drinking Water Policy, s 3(2).

The gist of the new scheme was to tone down the reform rhetoric and to demand only 'an element of token community contribution and involvement of user groups/panchayats in the selection and implementation of the schemes and for subsequent [operation and maintenance]'.[136] The government also indicated that it was ready to exempt communities from making a contribution in 'exceptional cases of hardship'.[137] Further, communities could choose the mode of contribution. This removed the previous insistence on a cash contribution.[138]

One explanation for these suggested changes was the fact that Swajaldhara had failed at two different levels. Firstly, the demand-led mode has not functioned as hoped by its promoters. This is in part due to the fact that they have not been able to create enough 'demand'. Indeed, where people have adequate access to water, they are loath to adopt the proposed reforms. Thus, it is in the areas where there is least availability of water that people are most responsive to the new scheme. This explains, for instance, why the pilot areas chosen for the Swajal project were areas facing tremendous drinking water problems. Secondly, there has been resistance from within the government to the new schemes. These have been perceived as eroding the power of existing departments. Thus, Swajaldhara was, for instance, disliked because it seemed to make the technical expertise of government departments irrelevant since their services are not necessarily required on a compulsory basis in the choice and design of schemes.[139] One of the new proposed features in reaction to this was that a Junior Engineer would be specifically charged with providing three to four gram panchayats technical assistance.[140]

This rethinking of the reform principles has eventually not been carried forward. This is illustrated by the fact that the commitment to demand-led reforms during the eleventh plan is more or less intact.[141] Indeed, the government has not shown any inclination to abandon the reforms at this stage. The idea is in fact to ensure that further reform measures are taken during the eleventh plan. The concept of a minimum community contribution of 10 per cent is, for instance, reasserted as something that needs to be part of all drinking water supply schemes.[142] Similarly, the operation and maintenance of all new single village schemes is to be borne by the community and further states governments are to progressively transfer all existing schemes to gram panchayats.[143] This has now been translated in instructions from the Union government that seek to carry Swajaldhara forward as originally defined. In other words, the government is

[136] Rural Water Supply and Sanitation, Eleventh Five Year Plan—Approach Paper (2006) 7.
[137] ibid.
[138] ibid.
[139] Remarks of RM Tripathi, Executive Engineer, Jal Nigam 29 August 2007.
[140] Rural Water Supply and Sanitation, Eleventh Five Year Plan—Approach Paper (2006) 7.
[141] Planning Commission—Government of India, *Eleventh Five Year Plan 2007–12—Volume II—Social Sector* (New Delhi: Oxford University Press, 2008) 163.
[142] ibid 48.
[143] ibid.

back to suggesting that 20 per cent of projects should be implemented under Swajaldhara principles and these projects are financed entirely by the Union government rather than shared between a state and the Union.[144] This does indicate that the time is not yet ripe for mainstreaming Swajaldhara to all ARWSP projects but also shows that the commitment to a community contribution is maintained. Indeed, the report of a meeting of State Secretaries on rural drinking water states that from 2008 onwards, 'the Swajaldhara scheme is going to be expedited drastically'.[145]

This happens to coincide with a similar commitment to the principles of the Swajal project by the World Bank. Thus, the Uttaranchal Rural Water Supply and Sanitation Project is, for instance, premised on the 'success of demand-driven community participatory approach'.[146] The Bank anticipates that the current project can contribute to the replication of the model in other states and the mainstreaming of the approach by the Union government that the Bank could then support.[147] This project seeks to go beyond Swajal in strengthening the involvement of the panchayati raj institutions whose limited involvement in the earlier scheme were seen as detrimental to the sustainability of the schemes.[148] In the context of this World Bank project, the cost-sharing principle has not been abandoned and communities will bear 10 per cent of the cost of all new investments. The only relaxation is that only 2 per cent of the cost must be in the form of cash while the other 8 per cent can be in the form of labour or cash.[149]

C. Drinking Water and the Realization of the Human Right to Water

The analysis carried out in earlier sections of this chapter highlights the contrast between the drinking water policies followed until the 1990s and the reforms that have been put in place since then. The changes that have been introduced raise broader questions concerning the role of the present reforms in the realization of the human right to water. This section examines the ways in which pre- and post-reform policies contribute to the realization of the human right to water. Issues related to some other elements of the human right to water not specifically addressed in this chapter are then taken up in the next chapter.

[144] Rajiv Gandhi National Drinking Water Mission, Allocation of funds under Accelerated Rural Water Supply Programme (ARWSP) during 2008–09, No. G–11011/5/2008–DWS.I (2008).

[145] Minutes of the State Secretaries' Conference on Rural Drinking Water and Sanitation held on 13 and 14 May 2008, Surajkund, Haryana, 4.

[146] World Bank, India—Uttaranchal Rural Water Supply and Sanitation Project, Project Appraisal Document (Report No. 35464-IN, 2006) 2.

[147] ibid 3.

[148] ibid 7.

[149] ibid 14.

With regard to pre-reform policies, it is striking that at least since the late 1960s, the government has operated schemes on the basis that water had to be provided free to all the inhabitants of rural areas. The premise was the state's duty to ensure that citizens have access to sufficient drinking water. This policy has operated until recently and has been the basis for the spread of rural drinking water schemes throughout the country. In fact, even a draft memorandum of understanding between the Union and the states for the eleventh plan states that there is a constitutional obligation to provide access to safe drinking water to the rural population.[150] This is in keeping with the case law, which recognises that the state has the responsibility to provide unpolluted drinking water.[151]

This policy framework has contributed to the introduction of millions of handpumps in Indian villages. The policies of the Union and state governments have brought immense relief to millions of people throughout the country. At the same time, there remain vast gaps in coverage and one of the weakest points of governmental action has been the lack of effective maintenance of existing schemes.

Overall, the government has acted under the perception it was its duty to provide free water to its citizens. This is of the utmost importance in a context where the human right to water is recognized as a fundamental right and a human right. Indeed, the fact that governmental policies have over the past several decades repeatedly emphasized the need to implement free water policies indicates that this has been the government's response to the judicial recognition of the human right to water. The fact that this policy of free water has not been applied to all persons in the country does not contradict this conclusion. Indeed, the rural population has been and remains by far the largest segment of the population and in overall numbers rural poverty dramatically overshadows urban poverty. The government has thus clearly recognized the need to focus on the specific situation of rural India. This is in fact confirmed in an approach paper for the eleventh plan that specifically recalls that the provision of safe drinking water to all facilitates the realization of the fundamental right to life of which clean water is a component.[152] The government also justifiably started by focusing on the part of the country that suffered more from poverty. Similarly, the government has acknowledged that unserved villages should be given priority over villages where supply is already provided as per the minimum norms in force.[153] This indicates a concern to ensure the realization of the minimum core content of the human right to water for all before providing enhanced coverage to people already provided with water.

[150] Memorandum of Understanding Between State Government of __ and the Department of Drinking Water Supply, Ministry of Rural Development, Government of India (2007) s 2.

[151] *Hamid Khan v State of Madhya Pradesh* AIR 1997 MP 191 (High Court of Madhya Pradesh, 1996).

[152] Rural Water Supply and Sanitation, Eleventh Five Year Plan—Approach Paper (2006) 1.

[153] ibid 5.

The starting point is thus that the government has sought to implement the human right to water to the best of its ability. It falls short of full realization but there has been progressive implementation. This is in keeping with the stipulation of the International Covenant on Economic, Social and Cultural Rights (ICESCR) that imposes on member countries to take measures that contribute to the realization of the rights over time.[154] There is no doubt that the Indian government has felt under an obligation to realize the human right to water even if there is no legislation that specifically makes the link, something that India shares with many other countries.[155] The implication is that the human right to water was deemed so fundamental that the question of its status never actually arose. The importance of drinking water has in fact been long recognized in legislation. The Bengal Irrigation Act 1876 recognized, for instance, that where irrigation works led to a substantial deterioration or diminution of drinking water supply, the government had a duty to provide an alternative supply of drinking water 'within convenient distance'.[156]

The implicit implementation of the human right to water takes added importance in the context of the reforms that are being introduced in the provision of drinking water in rural areas. Indeed, it is questionable whether the new policy framework that is being introduced constitutes a set of measures that further contributes to the realization of the human right to water or threatens some of the gains made earlier.

The new policies being introduced in India and in other countries of the South for the past decade seek to take these countries on a completely different path that conceives water as an economic good, that contemplates imposing on each individual community an increasingly important burden of their own water supply and that generally conceives of a reduced role for the government and a concomitant increase of the role of private sector actors in delivering drinking water.

The implementation of these new policies will lead to outcomes that are at least in some cases unacceptable from the point of view of established measures of equity and will directly or indirectly lead to violations of the human right to water. While the demand-led paradigm benefits a segment of the rural population, it affects the poorest by bypassing them, it creates increased inequalities in access to water and in the long run the imposition of operation and maintenance costs to each village individually will lead to reduced access to water in villages less well-endowed with water. Such policies need to be reversed because water is far too fundamental for human life. The imposition of operation and maintenance

[154] International Covenant on Economic, Social and Cultural Rights, New York, 16 December 1966, 993 UNTS 3 (1976) art 2(1).

[155] United Nations High Commissioner for Human Rights, Report on the Scope and Content of the Relevant Human Rights Obligations Related to Equitable Access to Safe Drinking Water and Sanitation under International Human Rights Instruments, UN Doc. A/HRC/6/3 (2007) 18.

[156] Bengal Irrigation Act 1876, s 12.

costs on rural communities does not seem to be based on rational justifications. As indicated in documents for the eleventh plan, one of the major policy concerns is that high operation and maintenance costs lead to the closure of water supply schemes because the government is not in a position to maintain and monitor assets. It is estimated that only 20 per cent of the required funds for operation and maintenance are available at present.[157] In a situation where the government is unable to muster the necessary resources for operation and maintenance, it is highly unlikely that rural communities will be able to take on the job and do better than the government.[158] If at all they do better than the government, it will be out of desperation because nobody can survive without water. The implication will be that other vital needs will suffer since this will likely imply a transfer of resources within already tight budgets.

The fact that new policies seek to impose a new burden on local people on the assumption that they can do better than the government, including with regard to the capacity to raise funds is problematic. Indeed, if access to water becomes dependent mainly on the financial capacity of specific individuals or specific panchayats, the likelihood is that the poorest individuals and communities will be the least well served. The potential for such results is made more problematic by the policy basis on which such changes are introduced. Indeed, already more than a decade ago the World Bank acknowledged that willingness to pay would be dependent on the availability of alternative and traditional sources of drinking water as well as the quality and level of service being provided before the introduction of the reform.[159] The fact that such reforms would not necessarily be willingly accepted by people suffering from insufficient access to water as a rational choice was highlighted by the Bank's call for a widespread campaign 'to communicate the message that water is a scarce resource and must be managed as an economic good'.[160]

The issue of willingness to pay remains a central basis for ongoing and proposed reforms. Indeed, it provides the justification for the introduction of cost recovery principles for all uses of water and for all individuals. At the international level, the figures proposed as appropriate in terms of willingness to pay range between 3 and 5 per cent of income.[161] The widespread acceptability of such figures is doubtful because the middle classes do not usually pay such high percentages of their income for water. Thus, in Mexico the rich spend only 0.8 per cent of income on water while the poor spend 5.2 per cent. In the UK, the

[157] Rural Water Supply and Sanitation, Eleventh Five Year Plan—Approach Paper (2006) 5.

[158] Concerning the paucity of revenues raised by panchayats, eg M Govinda Rao & UA Vasanth Rao, 'Expanding the Resource Base of Panchayats—Augmenting Own Revenues' (2008) 43/4 *Economic & Political Weekly* 54.

[159] World Bank, India—Water Resources Management Sector Review—Rural Water Supply and Sanitation Report (Report No. 18323, 1998) 49.

[160] ibid 53.

[161] N Prasad, 'Privatisation Results: Private Sector—Participation in Water Services After 15 Years' (2006) 24/6 *Development Policy Rev* 669, 676.

poorest pay over 10 per cent of their income. Even in a less extreme situation like Italy, while the average expenditure for water is 0.84 per cent of the average family income, low-income families spend 1.54 per cent and the poorest spend 3.4 per cent of their income.[162] In this context, a recent publication that seeks to revive the stalled water reform project in Delhi is noteworthy. It finds that the average willingness to pay is Rs 215 for the approximate 8,000 households surveyed.[163] This means that the average household is willing to pay about Rs 215 more than the current average bill of Rs 140. This publication finds these results logical because they find that the average cost for households to cope with the limitations of the existing water supply system is Rs 187.[164] This does not tell the whole story. Firstly, the figures show that the poor do not invest in borewells (average cost Rs 17,000) and that the percentage of households investing in water filters (average cost Rs 5,400) rapidly diminishes along socio-economic lines.[165] The average coping cost is thus brought up by the much higher spending of the higher middle class on coping costs. Secondly, Delhi is a city where the average per capita income is Rs 5,561 and the existing water bill amounts to about 2.5 per cent of average incomes. The figure of Rs 215 proposed by Misra and Goldar as the average willingness to pay represents 3.86 per cent of average incomes. In other words, the article suggests that people are generally willing to pay 6.38 per cent of their income for water. This is a wrong basis to address what is the most fundamental vital need and one which is in any case one of the major concerns of the residents of Delhi throughout the year and becomes the priority number one for millions of people during the long summer months. Similarly, a recent World Bank paper discussing willingness to pay in rural areas in India rejects the methodology used elsewhere and specifically excludes from the computation households who do not pay for water today because this leads 'to an underestimation of affordability'.[166] While this may be the case from an economic point of view, this implies that the situation of many of the poorest people is not even within the purview of the analysis undertaken that is then used as a basis for law and policy measures. It is thus impossible to conclude as the World Bank does that 'the charges being collected at present are generally much lower than what the households can afford and are willing to pay'.[167]

It is also noteworthy that the proposed Delhi reform scheme is from the outset unaffordable according to the World Bank's own criteria. Indeed, the

[162] C Armeni, The Right to Water in Italy (Geneva: International Environmental Law Research Centre, IELRC Briefing Paper 1, 2008) 5 available at <http://www.ielrc.org/content/f0801.pdf>.

[163] S Misra & B Goldar, 'Likely Impact of Reforming Water Supply and Sewerage Services in Delhi' (2008) 43/43 *Economic & Political Weekly* 57.

[164] ibid 64.

[165] ibid 59–60.

[166] World Bank, Rural Water Supply in India—Willingness of Households to Pay for Improved Services and Affordability (Policy Paper 44790, 2008) 6.

[167] ibid 8.

1998 water sector review specifically indicated that water and sanitation services are affordable if the cost falls within 3 per cent of incomes.[168] If the response of policy makers to the latter point is to follow the same World Bank report and tailor water supply services, this would mean downgrading the infrastructure provided to the poor to ensure it fits within the affordability bracket.[169] Neither option is acceptable from either an equity or human right point of view.

There are additional concerns with regard to the realization of the human right to water. Indeed, as indicated above, the government has for decades attempted to foster the realization of the human right to water, for instance, by providing free drinking water infrastructure in thousands of villages. Over time, the government has shown itself to be concerned with the realization of the human right to water by repeatedly seeking to provide all villages with minimum levels of drinking water infrastructure. Numerous policy documents over the past few decades indicate that the government has seen drinking water provision as a central part of its overall mission. It has acknowledged a duty to provide at least 40 lpcd free to all villagers.

This commitment becomes significant in a context where the government is withdrawing from the provision of drinking water infrastructure. Indeed, as a party to the ICESCR, the government is bound to progressively implement human rights. The progressive withdrawal of the government from the provision of water supply services in rural areas and the imposition of a growing share of the capital and maintenance costs of drinking water to villagers constitute in all likelihood backward steps. They may thus lead to a violation of the commitments of the government under the ICESCR. This is illustrated by some of the examples above. The case of Jogela village highlights that local power brokers cannot be expected to perform better than the government in terms of ensuring priority access to the most needy villagers.[170] Similarly, in the case of people who rely on water sources built and maintained by the government, their inability to pay for better access today does not affect their often insufficient access today. It may, however, lead to a complete lack of access to water once the sources they use today become unusable unless they pay, whether they can afford or not. Reform policies thus need to be reconsidered in the light of the human right to water.

[168] World Bank, India—Water Resources Management Sector Review—Rural Water Supply and Sanitation Report (Report No. 18323, 1998) 49.

[169] ibid.

[170] See p 163 above.

6

Towards an Alternative Framework for Water Law Reforms

Water law is at an important juncture in India and in a number of other countries. A process of water legislation change has started and the process itself is welcome in view of the often antiquated rules and principles found, for instance, in laws dating back to the colonial era. While changes are desirable and necessary, the direction of the changes is something that needs further investigation. Firstly, the reforms that are being implemented in India and in many developing countries proceed from an international policy-making consensus that has never been effectively debated at the village, block, district, or state levels. In this sense, whether the reforms are the reforms that would be introduced if they had been introduced following the democratic governance structures established in India is immaterial. Ongoing reforms have not been subject to sufficient democratic scrutiny, something that needs to happen urgently. This is the case at the level of individual countries like India as well as at the international level even if some commentators satisfy themselves with the fact that the basic principles of water sector reforms 'are endorsed by nearly all water academics, professionals, and institutions including the World Bank'.[1] Secondly, as highlighted earlier, ongoing reforms are relatively narrowly conceived and in any case constitute only one of various potential reform avenues that could be pursued. This, together with some of the shortcomings of ongoing reforms highlighted earlier calls for further work on conceiving water law reforms that will not only foster better management in the water sector but more importantly contribute to the realization of the basic principles of water law, environmental law and human rights found in international law instruments, the Constitution, judgments as well as relevant legislation.

This chapter seeks to contribute to further thinking about water law reforms by proposing alternative bases for a series of comprehensive water law reforms. This

[1] H Ingram, JM Whiteley & R Perry, 'The Importance of Equity and the Limits of Efficiency in Water Resources', in JM Whiteley, H Ingram & R Perry eds, *Water, Place, and Equity* (Cambridge, Mass: MIT Press, 2008) 1, 2.

constitutes only one of several possible alternatives to ongoing reforms. The proposals are based on the understanding that water law needs to be conceived in such a way that it contributes to the realization of the human right to water, provides a socially equitable and environmentally sustainable framework and is based around an understanding of water and its links with the other areas of law that are directly or indirectly affected or concerned by the measures that may be taken in water-specific legislation.

This translates into two related priorities. Firstly, it is imperative to rethink water law so that its human rights, social, and environmental aspects get the priority they deserve, not only in general statements of courts or non-binding policy statements but in the actual legal framework that can make a difference to people's lives. In a context where the economic dimension of water is often given a priority, this implies taking a significantly different direction in policy terms. Secondly, to avoid some of the concerns that have arisen in the context of sectoral water sector reforms being implemented, the adoption of a water act that brings together all the basic principles of water law and makes them applicable throughout the sector is necessary. This can be undertaken at the state level by each individual state or at the Union level if the government musters the political will to do so. Regardless of the level at which this is proposed, a framework water law needs to be introduced. The need for a framework legislation is reinforced by the fact that international water law is currently insufficiently developed to provide the framework that is missing at the national level. Indeed, international water law still largely focuses on international watercourses and does not include a comprehensive framework setting out the basic principles of water law in the same way that the Rio Declaration can be said to provide a catalogue of international environmental law principles.[2]

A. Beyond Existing Reforms

Reforms of the existing water regulatory framework are necessary to update water law to face the challenges of the twenty-first century. However, proposed changes are neither the only possible answer to identified problems nor necessarily the best. This calls for a broader discussion of the possible alternatives that different Indian states and different countries in the South can choose in their attempts to update their water law. This preliminary discussion on options for reforms is a precondition for an informed, transparent, and democratic choice. One of the main problems with existing law reforms in India is not so much that they adopt one specific reform track but the fact that alternatives are not discussed and put on the table. As a result, voters, office holders at the local level, members of legislative

[2] Rio Declaration on Environment and Development, Rio de Janeiro, 14 June 1992, UN Doc. A/CONF.151/26/Rev.l (Vol. l).

assemblies or MPs are often only given a choice between retaining the status quo and adopting the reforms proposed. Since most people agree, though often for different reasons, that the status quo is not an effective response to existing challenges, the proposed reforms are adopted. In a democratic society where panchayats have been given relatively important powers over water and where access to water for life and livelihood is something that is of primary importance to each and every individual, this is insufficient in terms of process.

As indicated in earlier chapters, the present reform model is selective in its emphasis on certain problems and on certain solutions to address them. The principles proposed, largely derived in the first place from the Dublin Statement fail to take into account the intrinsic link between water and the realization of the human right to water and a series of other fundamental rights. Further they fail to take into account the essentially different nature of water as the primary source of life on earth after air for humankind and nature. What is in fact needed is a stronger recognition of the multi-faceted dimensions of water as a cornerstone of human survival, of the survival of all animals, and as a basic element contributing to meeting our food needs, irrigation, and energy needs and economic development.

The advantage of the proposed reforms is that they propose a relatively simple set of principles that can be used to rally people in different areas of the water sector. Alternatives that take a broader view of water law are likely to be more complex. Yet, there is nothing to gain by artificially simplifying complex problems. Water and its multiple dimensions and multiple use cannot be reduced to a simple set of issues. Thus, even if putting the human rights and environmental dimensions at the centre of the legal framework makes the system more complex, this only reflects the need for a comprehensive framework and one that is adapted to each specific context, such as the different states in India.

1. Limitation of existing water law and proposed reforms

The first reason for reforming water law is that the existing framework has more than outlived its utility. Thus, the framework provided by a number of irrigation acts, whether adopted before or after independence is based on the framework proposed during colonial times. Whether this was appropriate or not for the needs of the country at the time is not at issue here because regardless of the answer, the conceptual framework for these laws dates back to the late nineteenth century when issues in the water sector were drastically different. The property rights-based framework for accessing water has also been acknowledged as unsuited to present challenges.[3] The same is true of the groundwater model bill, which is the brainchild of late 1960s thinking but fails to reflect major advances in the field of human rights and environmental law as it has not been fundamentally rethought since 1970.

[3] National Commission for Integrated Water Resource Development Plan, Report (New Delhi: Ministry of Water Resources, 1999) xii.

In addition to the dated nature of existing water law, it is also unclear and difficult to understand. This is due in part to the multi-layered framework that is the hallmark of existing water law. This is apparent in different areas. With regard to control over surface water, it is relatively difficult to comprehend the legal framework. On the one hand, it is often agreed that the state has either the power to use water for public purposes or an absolute right to control it. This position is, however, largely derived from irrigation acts, which are the acts having taken a position on the matter. Given that water law has developed in a sectoral manner, it is arguable that the legislature did not consider other water uses at the time of the adoption of these acts. On the other hand, the evolution of water law principles as adopted by the Supreme Court is not necessarily reflected in the existing legislation.[4] Thus, none of the laws adopted in the past decade have addressed the basic question of control over water or the fundamental right to water. This results in an increasingly asymmetric legal framework where old principles coexist with new directives of the Supreme Court and where recent laws do not seem to take notice of either.

Proposed reforms seek to address some of the problems identified in the water sector. However, the first limitation of ongoing water law reforms is that the starting point is the assessment done in the context of water sector reforms. This has two implications. Firstly, water sector reforms do not specifically focus on legal aspects and, as identified earlier, reforms are mostly based on economic principles. This is, for instance, illustrated by the Uttar Pradesh water policy. The latter specifically calls for legal intervention in a number of areas. All these interventions are to be conceived within a framework that seeks to ensure the 'most efficient use' of water.[5]

Secondly, law reforms fail to effectively build on the existing legal framework. Doing so would imply a comprehensive review of existing legal and institutional arrangements to determine what needs to be kept, updated or changed in accordance with updated circumstances in law—such as constitutional amendments or Supreme Court judgments—and in the water sector in general. The absence of such comprehensive exercise leads to the addition of new layers of rights and obligations that are superimposed on existing frameworks. This may be politically convenient because it saves the government the need to bring legislation that specifically amends existing acts and regulations but leads to a situation where certain rights and arrangements become extinct without formally acknowledging it. It also leads to the adoption of arrangements that have the potential to conflict with Supreme Court decisions.

Ongoing water law reforms also fail to take into account lessons from history. This is particularly visible in the context of drinking water. From the second half of the nineteenth century, in European and other countries local authorities

[4] On the public trust, ch 2.B.2.
[5] Uttar Pradesh Water Policy 1999, s 17(1).

found themselves forced to buy water companies to make provision of drinking water universal since private water companies often made initial investments that were too small, neglected maintenance and showed no interest in providing water supply to poor households.[6] While health concerns may have been determining factors at the time, other concerns, such as the fulfilment of the human right to water have come to predominate in recent times. Apart from some countries such as France that never fully nationalized its drinking water sector and England and Wales that privatized their water sector in the late 1980s, these countries have never gone back on their commitment. The situation of the Canton of Geneva in Switzerland illustrates this point well. While the canton had had a mixed system of a public utility serving most people and smaller private utility serving a significant minority of the population for most of the twentieth century, drinking water was eventually made entirely public in the 1980s because the private utility failed to provide the same level of service as the public utility.[7] Reforms that simply seek to blame the state for failing to provide water to all thus make short shrift of historical evidence showing that private utilities are not well placed to provide good service to the poor. Rather than insisting on private sector participation, reforms could seek ways and means to ensure better delivery through the public sector or through decentralized bodies of governance in novel ways that recognize the character of water as essentially beyond appropriation.[8]

Additionally, ongoing reforms do little more than pay lip service to the need for solutions tailored to local circumstances. While policy documents for water sector reforms repeatedly state that no one model is relevant throughout the world and throughout India,[9] the practice of water law reforms does not demonstrate this. On the one hand, there is significant variety among the different states of India that mirrors the diversity of situations between countries around the world. This is true in terms of environmental conditions since different states experience conditions ranging from extremely dry and hot and extremely dry and humid to humid cold and dry cold climates. This is also true in social and economic terms since socio-economic conditions differ significantly in different parts of the country. On the other hand, the reality is that the various state water policies could be relatively easily amalgamated into a common document. Similarly, states adopt acts that are copycat of each other. This is the case of water user association

[6] C Ward, *Reflected in Water—A Crisis of Social Responsibility* (London: Cassell, 1996) 96 and Ingram, Whiteley & Perry (n 1 above) 7.

[7] M Ruetschi, Déprivatisation de l'eau—l'expérience du Canton de Genève (Geneva: International Environmental Law Research Centre, IELRC Briefing Paper 2008–03, 2008) available at <http://www.ielrc.org/content/f0803.pdf>.

[8] Solutions exist as exemplified, for instance, in the case of the city of Porto Alegre. H Maltz, 'Porto Alegre's Water: Public and for All' in B Balanyá et al. (eds), *Reclaiming Public Water—Achievements, Struggles and Visions from Around the World* (Amsterdam: Transnational Institute and Corporate Europe Observatory, 2nd ed. 2005) 29.

[9] World Bank, India—Water Resources Management Sector Review—Inter-sectoral Water Allocation, Planning and Management (Report No. 18322, 1998) 52.

legislation that looks alike in most states or the case of legislation that is simply lifted from that of another state a couple of thousand kilometres away like in the case of Arunachal's water regulatory legislation.[10] This is surprising in a context where even the UNDP makes the point that there are no ready-made solutions that can be successfully implemented in all countries of the world.[11]

2. Alternative bases for law reforms

One of the biggest shortcomings of water law is that it has never yet been broadly conceived. 'Old' water law principles and legislation focused in large part on economic development concerns. 'Reformed' water law focuses in large part on economic priorities and management concerns. Neither can claim to constitute a broad framework that encompasses all the dimensions of water. In particular neither has effectively put at the centre of water law human rights, livelihood concerns, and ecosystem needs. A broader water law needs to be based on a more diverse set of bases than is the case at present.

Firstly, poverty eradication must be the main focus for the development of water law. While poverty reduction or eradication is central in terms of development policy, a lot more needs to be done to ensure that water laws effectively put these objectives at the centre of their mandate. This is illustrated by the existence of policy documents that blame the water crisis on the poor, such as a preparatory document for UNCED that identifies two main causes for the degradation of water resources, poverty, and economic activity, where poverty is put in the first place.[12] In fact, it is not the poor that are the major cause of environmental degradation but the poor that disproportionately suffer from degradation whose benefits they have not even enjoyed.[13] As a result, policies premised on blaming the poor as a cause of the problem are intrinsically flawed. In the case of water, for instance, it is not the poor that are the cause of increasing water scarcity, since it is always the poorest that have the least access to water. It is also not the poorest that contribute to water pollution since they are neither landowners contributing to water resource degradation through irrigation nor industrialists polluting water as part of their economic activities. In fact, it is the poor that nearly always suffer. As a result, the law and policy framework needs to be built around measures that contribute to lessen the disadvantages and discrimination that the poor face rather than around measures blaming them for the situation they find themselves in.

[10] See p 119.

[11] United Nations Development Programme, *Human Development Report 2006—Beyond Scarcity: Power, Poverty and the Global Water Crisis* (New York: UNDP, 2006) 11.

[12] Preparatory Committee for the UNCED, Protection of the Quality and Supply of Freshwater Resources: Application of Integrated Approaches to the Development, Management and Use of Water Resources, UN Doc. A/CONF.151/PC/73 (1991) 8.

[13] United Nations Environment Programme, *Global Environment Outlook—GEO4—Environment for Development* (Nairobi: UNEP, 2007) 315.

Secondly, water law needs to be built around India's development needs.[14] These needs cannot, however, be equated with economic development. From the perspective of water law, development needs that should constitute the bedrock of new frameworks are primarily those of the majority of people who have not benefited from economic growth. This also includes the need for a strategy of development that is both socially equitable and environmentally sustainable. The development concerns that need to inform water law and policy are in the first place India's. This assumes particular importance in a context where international water policy instruments play a significant role in shaping up national water laws. As highlighted earlier, the influence of the international consensus over water policy is quite significant. This does not, however, imply that these are the best measures for each and every country or each and every Indian state. Regardless of the assessment that is made of ongoing water sector reforms and water law reforms, there is a need for a law making process that starts from local democratic governance bodies rather than from international policy making bodies.[15]

Thirdly, new water laws need to be based on a set of broad principles that cover the whole water sector. An 'integrated' perspective is thus appropriate. This should, however, be different from the concept of integrated water resources management (IWRM) because the integrated conception must go beyond a focus on management. While IWRM correctly calls for a broader view that, for instance, considers water at the river basin level, its conceptual framework is largely focused on the need for more efficient water use. What is in reality needed is more efficiency as part of a broader package of measures that starts by looking at people's needs and environmental needs. This broader perspective requires the integration of the needs of people for the fulfilment of the human right to water and livelihood options together with ecosystem needs, irrigation needs, and industrial development needs. In this scheme, it is not efficiency that predominates but rather equity, human rights, and ecology. This leads to different policy prescriptions than what is achieved through IWRM. Thus, the primary question is not whether and how to transfer water to water deficit basins but rather how to allocate water for non-essential activities after having provided for the fulfilment of the human right to water as well as ecosystem needs. Since under most circumstances, there is more than enough water for these basic functions, they need to be given first priority before other water uses are considered. Where water is insufficient to cover all existing water uses, 'integrated' water law means that it needs to be considered in conjunction, for instance, with agricultural policies. Where a region cannot sustain the existing level of irrigation, incentives to grow different cash crops may be one of the first answers to the policy questions that arise.

[14] This was already the assessment made by C Singh, *Water Rights and Principles of Water Resources Management* (Bombay: Tripathi, 1991).

[15] Concerning the problems posed by standards defined at the international level, eg DP Fidler, 'A Kinder, Gentler System of Capitulations? International Law, Structural Adjustment Policies, and the Standard of Liberal, Globalized Civilization' (2000) 35 *Texas Intl L J* 387.

Fourthly, as suggested by existing reforms, water scarcity needs to inform water law making for the future. Indeed, relative—and in some cases absolute—water scarcity already exists in some places. Yet, there are different ways in which water scarcity can be conceived and used in policy terms. At present, water scarcity is often seen in terms of physical scarcity and blamed, for instance, on population growth that is seen as a central cause of diminishing per capita availability of water.[16] While population growth is a central concern of development policy it cannot be the primary rationale for water law reforms. Indeed, while physical scarcity is important, social aspects are at least as significant from a law making perspective. Additionally, where scarcity exists, for instance, because the quality of water is insufficient to be drunk by humans, this is often caused by other human activities such as agriculture which is responsible for 70 per cent of water pollution on a global level as well as industry and urban life.[17] Thus, scarcity cannot be considered in a vacuum in the development of water laws. Overall, while physical scarcity is important, issues such as access to water, prioritization of use, and environmental concerns should be given much more weight in law making.

Fifthly, the role of the various actors in the water sector must be seen in a much more nuanced way than what water sector reforms have done over the past few years. With regard to the state, while it has failed to provide all the benefits that could be expected from a welfare state, the trend has been much more positive since independence than in the previous period. In the context of access to drinking water in particular, government interventions have made tremendous inroads towards the realization of the human right to water over the past couple of decades. In addition, government intervention has, despite its shortcomings, been in many cases the only type of intervention that actually benefits the poorest and most disadvantaged. Since the fulfilment of human rights needs to be looked at from the perspective of the impact of policies on the poorest, the government has on the whole done at least a commendable job. The shortcomings of government intervention abound and have been discussed for many years. They must indeed be addressed to ensure, for instance, that government intervention is much better targeted towards the poorest and that more of the money spent actually reaches the people who are supposed to benefit. These shortcomings notwithstanding, the state remains the only actor that can effectively foster the realization of the human right to water. New ways to make state institutions more accountable to the people they are supposed to serve need to be found. A combination of measures can be proposed including, for instance, more decentralization of the kind

[16] Preparatory Committee for the UNCED, Protection of the Quality and Supply of Freshwater Resources: Application of Integrated Approaches to the Development, Management and Use of Water Resources, UN Doc. A/CONF.151/PC/73 (1991) 2.

[17] BR Johnston, 'The Commodification of Water and the Human Dimensions of Manufactured Scarcity' in L Whiteford & S Whiteford (eds), *Globalization, Water, and Health—Resource Management in Times of Scarcity* (Santa Fe: School of American Research Press, 2005) 133.

initiated under the 73rd and 74th constitutional amendments,[18] better access to information through a stronger right to information scheme, and the strengthening of schemes like the one proposed in the National Rural Employment Guarantee Act 2005 that have the potential to address poverty and marginalization in ways that give people much more control over their destinies.[19] In other words, what is required is not plans to shrink the government's powers by turning it from a provider to a facilitator, but rather to find new ways in which the government can deliver the social and environmental benefits it is supposed to deliver. This is necessary because the private sector can never effectively step into the government's shoes to ensure the realization of the human right to water.

B. New Framework for Access to and Control over Water

The question of control over water and access to water has been at the centre of most water law in the modern era. In other words, what has been at the centre of debates is the question of property rights over water. As analysed earlier, the response in India from the late nineteenth century to the late twentieth century has been to recognize the state's predominant control over water under its jurisdiction. Additionally, rights of access to water at the individual level were and still are largely linked to land rights.

The connection between control over land and access to water has been under strain. Indeed, it does not reflect the long-standing understanding that water cannot be appropriated. This has been challenged from two different positions. On the one hand, the Supreme Court has established the non-proprietary nature of water by declaring surface waters a public trust. This is significant because it severs the property right connection between the state and water and re-establishes the special nature of water as a special substance that is beyond state or individual appropriation because of its fundamental nature for human life and ecosystems. On the other hand, the link between individual land rights and control of water has been challenged in ongoing water sector reforms because it does not allow for the trading of water entitlements separately from the land itself. Indeed, the MWRRA has for the first time set up a system that allows trading of water entitlements without reference to land rights. As opposed to the Supreme Court proposal, this proposes to strengthen individual property rights over water.

[18] One of the shortcomings of the constitutional amendments is that they need to be translated at the state level and power needs to be effectively delegated, something which has been unevenly realised. cf MS Vani, 'Reviving Customary Law: Enabling Law for Water Harvesting' in A Agarwal, S Narain and I Khurana (eds), *Making Water Everybody's Business—Practice and Policy of Water Harvesting* (New Delhi, Centre for Science and Environment, 2001) 332.

[19] eg P Ambasta, PS Vijay Shankar & M Shah, 'Two Years of NREGA—The Road Ahead' (2008) 43/8 *Economic & Political Weekly* 41 and R Khera, 'Empowerment Guarantee Act' (2008) 43/35 *Economic & Political Weekly* 8.

1. From public trust to common heritage status

The first element that needs consideration is the legal status given to water. The recognition that water is a public trust is a big step forward in confirming the different legal status of water. The main point that still needs to be addressed in this regard is the need to include groundwater as part of the public trust since all water should be governed by the same principles.[20]

Nevertheless, the public trust doctrine is not the most appropriate basis for modern water law and further thinking needs to be developed in this area. Two main concepts developed in large part in the context of environmental law provide possible answers. The first is 'common concern of humankind', a notion that has arisen in the context of international environmental agreements. Common concern refers to issues that are of relevance to everyone involved, all states of the international community or all individuals at the national level. Its special contribution at the international level is to recognize that even where states refuse to part with their sovereign powers, there is a mid-way solution that preserves sovereign control but recognizes the need to cooperate in view of the global relevance of a specific issue. Two of the main cases where the principle of common concern has been accepted define its scope and relevance. In the context of global warming, common concern alludes to the global nature of the problem and the need to cooperate even though global warming touches on many issues that go back to the core of the notion of sovereignty.[21] In the context of biological diversity, common concern emphasizes the fact that even an issue which can in part be confined to the national sphere has global relevance because biodiversity conservation and use has impacts that go far beyond the national sphere.[22]

Water fits in both categories of common concern issues. Firstly, as already recognized in international water law, issues concerning transboundary watercourses are international in scope. Secondly, water is also eminently international in scope in its other dimensions. Even though the freshwater that sustains basic water uses such as domestic uses is mainly found within each state, water as a substance is much less amenable to physical demarcation than biodiversity. As a result, water is an easy candidate for common concern status. In fact, if states could agree on biodiversity being a common concern of humankind, water is in principle a much easier case.[23]

[20] cf T Shah, 'Groundwater Management and Ownership: Rejoinder' (2008) 43/17 *Economic & Political Weekly* 116, 118 arguing against because it will make no 'material difference'.

[21] United Nations Framework Convention on Climate Change, New York, 9 May 1992, UN Doc. A/AC.237/18, preamble.

[22] Convention on Biological Diversity, Rio de Janeiro, 5 June 1992, Doc. UNEP/CBD/94/1, preamble.

[23] This is not accepted by the majority of states and was in fact specifically debated during the negotiations for the 1997 Convention. On this point, eg PW Birnie & AE Boyle, *International Law and the Environment* (Oxford: Oxford University Press, 2002), 140.

While the notion of common concern has been developed at the international level, it is also relevant at the national level. It provides a reminder that even if control over water can in a number of situations be physically exercised, the larger ramifications of certain water uses must be looked at from a broad perspective. The case of groundwater illustrates the fact that even if landowners have direct control over water found under their land, this is partly illusory since today's pumping devices provide the means for drawing water from under a neighbour's land. In other words, even in a context where some form of individual control is allowed, it is imperative to take a broader view that recognizes the essentially different nature of water.

Common concern is conceptually easy to extend to water. Its application to water will, however, not lead to significant changes in practice. A much more significant paradigm shift is needed. A number of reasons call for a conceptual framework that leaves aside property rights altogether. Indeed, from the local to the international, legal concepts that seek to recognize rights of control by some actor or the other are by definition based on a misconceived understanding of the nature of water. From the local to the global level, the water that humankind uses is closely dependent on freshwater replenishment brought about by the global water cycle, itself directly related to global climatic patterns. There is, in fact, a direct link between global warming and freshwater availability since it will, for instance, cause many semi-arid areas to suffer a decrease in water. The Intergovernmental Panel on Climate Change finds with 'high confidence' that the negative impacts of climate change on freshwater systems outweigh the benefits.[24] The close relationship between the water that we actually use—as opposed to freshwater that is theoretically available in glaciers and polar areas or water that can be made available, for instance, through desalinization—and the global water cycle is an indication of our incapacity to control water. This global nature of water needs to be recognized in law as the basic principle on which water laws are built.[25] This is necessary because 'water resources of the world have been and always will be shared, as the hydrological cycle redistributes and renews resources'.[26] The same reasoning applies at the national level, as recognized, for instance, in South Africa.[27]

Among existing legal constructs, the principle of common heritage of humankind provides a good starting point to reconsider the legal nature of water. The idea of a common heritage of humankind was brought forward in the context of the negotiations for a new convention on the law of the sea in the late

[24] Intergovernmental Panel on Climate Change, Fourth Assessment Report—Climate Change 2007: Synthesis Report (Geneva: IPCC, 2007) 49.

[25] cf SC McCaffrey, *The Law of International Watercourses* (Oxford: Oxford University Press, 2007) 168 arguing that the international community has a vital interest in the global hydrological cycle.

[26] International Conference on Water and the Environment (Draft), Doc. HWR 320 (August 1990) s 1, para 4.

[27] South Africa, National Water Resource Strategy (2004) app A, principle 2.

1960s.[28] It serves to highlight that there are situations where neither the assertion of sovereignty nor the absence of an access regime offer a solution that leads to equitable and sustainable outcomes. In the case of the law of the sea negotiations, debates concerning common heritage focused on deep seabed minerals. Under older principles, deep seabed minerals were part of the high seas, an area where freedom of navigation was the key operative principle. Countries which did not have the technological capacity to exploit deep seabed minerals saw the principle of free appropriation underlying the freedom of the high seas as likely to lead to a situation where the already more powerful and technologically advanced states would reap a disproportionate share of the benefits. The idea of a common heritage status is to recognize that a resource that does not belong to anyone in particular should both be preserved and used for the benefits of all. The principle of common heritage of humankind was accepted by negotiators in the case of the UNCLOS adopted in 1982.[29] In fact, this became one of the major conceptual advances of the 1982 convention compared to the 1958 conventions.[30]

The concept of common heritage of humankind implies firstly that there can be no sovereign claims on a resource. It further recognizes the need for joint measures to protect and exploit the resource. This is to prevent overexploitation and unsustainable exploitation as well as to prevent disproportionate access by certain actors. In its most evolved form, the principle of common heritage applied at the international level implies that an international institution has the power to regulate the use and conservation of the resource under its jurisdiction. In the context of the law of the sea, this proved too far-reaching for some states and an amendment had to be adopted to ensure the entry into force of the convention.[31]

The principle of common heritage can be usefully applied to water.[32] Firstly, water found in a given location at a certain point in time is always transient. In fact, this has long been recognized and explains, in part, why jurists shied away from suggesting full property rights over water but rather opted for usufructuary rights.[33] Secondly, it is understood today that freshwater is part of a global cycle that no individual and no government has the capacity to control. Thirdly, water

[28] United Nations General Assembly Resolution 2749 (XXV), Declaration of Principles Governing the Sea-Bed and the Ocean Floor, and the Subsoil Thereof, beyond the Limits of National Jurisdiction, 17 December 1970, UN Doc. A/RES/2749 (XXV).

[29] Part XI, United Nations Convention on the Law of the Sea, Montego Bay, 10 December 1982, UN Doc. A/CONF.62/122.

[30] eg T Scovazzi, The Concept of Common Heritage of Mankind and the Resources of the Seabed Beyond the Limits of National Jurisdiction (Paper prepared for the international workshop on Resources of the Seabed and Subsoil, Buenos Aires, 15–17 May 2006).

[31] Agreement Relating to the Implementation of Part XI of the United Nations Convention on the Law of the Sea of 10 December 1982, New York, 28 July 1994, UN Doc. A/RES/48/263.

[32] cf R Petrella, *The Water Manifesto: Arguments for a World Water Contract* (London: Zed, 2001) 91 proposing that water should be recognised as a 'common global heritage'.

[33] D Guillet, 'Water Management Reforms, Farmer-Managed Irrigation Systems, and Food Security—The Spanish Experience' in L Whiteford & S Whiteford (eds), *Globalization, Water, and Health—Resource Management in Times of Scarcity* (Santa Fe: School of American Research Press, 2005) 185.

cannot be compared to any other substance because of its special role in sustaining life on earth. Fourthly, all water needs to be treated consistently, regardless of its source or use.[34]

On the whole, water availability is dependent on a chain of actions stretching from the local to the global level, none of which can be broken without having direct impacts on the others. Indeed, while rainwater harvesting and groundwater recharge in general are of great importance at the local level, this can only be usefully undertaken if groundwater is not polluted by other water uses. In turn, local actions are influenced not only by water laws and policies but also, for instance, by environmental laws and policies. Additionally, national water policies are influenced, for instance, by policy actions taken at the national and international levels concerning global warming. It is therefore unhelpful to conceive of water as a substance that can be controlled and appropriated. Rather, it should be conceived of as a common heritage that needs to be conserved and used in equitable and sustainable ways. The genealogy of the concept in the context of the law of the sea clearly indicates a focus on the weaker and less economically developed countries. It thus lends itself to easy use in the case of water since it incorporates a strong equity dimension both at the national and international levels.

One of the implications of the use of the common heritage of humankind is that issues related to access to water, water quality and water conservation are not an individual matter but a common task and responsibility. This implies that institutions that have as their core role the promotion of the broader public's interest and the poorer in particular must play a central role in the water sector. At the national level, the state is the constitutionally sanctioned representative of the broader interests of all individuals. Yet, the major difference with the notion of the public trust is that the state is not conceived any more as a trustee that can for all practical purposes in large part determine what are the best interests of the public, with limited accountability to the people on whose behalf it acts. In the case of common heritage of humankind, the state is in the position of an actor that is accountable for its actions. In practice, solutions need to be devised to ensure that there is effective accountability, one of the big challenges of the reforms that need to be undertaken in the water sector. At the international level, the application of the principle of common heritage is conceptually easier and practically much more difficult. Indeed, what common heritage status requires is an international institution that oversees water and has at least certain powers it can impose on states. Such restrictions on sovereignty are what states have found difficult to agree on in many cases. There is, however, no reason why new patterns cannot evolve in years to come. Indeed, while the physical inaccessibility of deep seabed minerals or resources on outer-space bodies made a consensus around the

[34] This is, for instance, recognised in South Africa, National Water Resource Strategy (2004) app A, principle 2. Also S Hodgson, Land and Water—The Rights Interface (Rome: FAO, FAO Legislative Study 84, 2004) 51 arguing that 'the concept of distinguishing private waters from public waters is something of a nonsense from a hydrological perspective'.

principle of common heritage possible, global warming, water scarcity and water quality problems may soon force states to rethink their basic positions. Once the nature of water as part of a global cycle that no state can influence individually is established and in a context where all known patterns of water flows are fast changing because of global warming,[35] political opinions are likely to change relatively swiftly.

Treating water as a common heritage of humankind provides a useful way to avoid the pitfalls of other comparable concepts such as 'public good'. The notion of public good provides an interesting starting point to think about the nature of water. Thus, it can, for instance, be extended to cover water supply, the obligation of providing water becoming the public good, rather than water itself.[36] It does not, however, provide the answers that are needed in law and policy. This is due to the fact that the public good status of water is open to debate because the definition implies that the good is available without restriction, something which does not obtain in many specific locations.[37] Additionally, the notion of public good is unhelpful to guide the development of basic legal principles guiding water law because it is an economic notion that cannot be directly translated as a legal principle. Finally, the emphasis on 'good' in public good makes it unsuitable for water since water in legal terms should not be considered as a good.

2. Delinked access to water from land rights

One of the consequences of the application of the principle of common heritage of humankind is that there can be no individual property rights over water. In principle, this takes care of the need to sever the link between control over land and access to water. While this is accurate, additional discussion is required because there are different ways to delink water from land rights.

As alluded to in chapter 2, the primary problem is that the link between access to water and land ownership neither provides the basis for socially equitable water policies nor for environmental sustainable water policies. The skewed patterns of land ownership and the fact that more than 40 per cent of the rural population is landless make land one of the worst possible starting points for the development of socially equitable policies concerning drinking water and access to water for livelihoods. It also fails to provide a basis for environmentally sound regulation of water, in particular groundwater since no individual landowner has a specific incentive to safeguard water, in particular in situations where s/he can draw additional water from under a neighbour's land with sufficiently powerful

[35] This includes fast receding glaciers, reduced precipitation in already dry areas and increased precipitation in other areas.

[36] J Boesen & PE Lauridsen, '(Fresh)water as a Human Rights and a Global Public Good' in EA Andersen and B Lindsnaes (eds), *Towards New Global Strategies: Public Goods and Human Rights* (Leiden: Martinus Nijhoff Publishers, 2007) 393.

[37] ibid 399.

pumping devices. While the problems linked to the nexus between groundwater and land ownership in particular are severe, not all actors accept yet the necessity of drastic changes. This is, for instance, the case of the Expert Group set up by the Planning Commission, which concludes that 'no change in basic... legal regime relating to groundwater seems necessary'.[38]

Different answers can be given to the problem identified. In the context of water sector reforms, delinking is proposed as a way to free water transactions from the limitations imposed by the link with land. In other words, delinking is proposed as part of measures to bolster water use efficiency. This is, for instance, what is being attempted in the context of the MWRRA with the setting up of tradable water entitlements conceived as rights that can be 'transferred, bartered, bought or sold on annual or seasonal, basis within a market system'.[39] In other words, under water sector reforms, there is a direct connection between delinking land rights and the introduction of new forms of private property rights over water. Yet, this is not as clear-cut as it appears at first sight. Indeed, land rights remain a major factor in assigning original tradable rights.[40] Delinking has been associated with significant problems in countries where it has already been implemented. Thus, in Spain turning water into a tradable commodity has had the unfortunate result of having a few big landowners capturing all or most of the new rights, thus creating new 'waterlords'.[41] Additionally, where there is no requirement for the land to which rights are attached to be in use, this rewards absentee landlords and land speculators.

Moving beyond water sector reforms, alternative proposals for delinking water and land focus on addressing the social and environmental shortcomings of the existing framework. This is something which remains novel because little has been done to make it a reality but it has been considered before as a general issue, for instance, by Chhatrapati Singh.[42] In a country influenced by common law principles like in India, a good starting point is to follow South Africa's lead in simply abrogating riparian rights.[43]

Delinking land and water should firstly lead to rethinking the nexus between the definition of a water 'user' and land in the context of irrigation. In other words, there is a need to rethink the way in which water user associations are conceived and the kinds of benefits they are supposed to deliver. Indeed, there is a strong presumption that a system of water allocation that excludes about 40 per cent of the population is discriminatory. With regard to groundwater, the challenge is to make sure that the source of water that fulfils most basic human

[38] Ground Water Management and Ownership—Report of the Expert Group (New Delhi: Government of India, Planning Commission, 2007) 41.

[39] Maharashtra Water Resources Regulatory Authority Act 2005, s 11(i).

[40] World Bank Website, Empowering Users by Giving Them Clear Water Entitlements, available at <http://go.worldbank.org/6TY7X9U3H0>.

[41] Guillet (n 33 above) 189.

[42] Singh (n 14 above) 39.

[43] South Africa, National Water Resource Strategy (2004) app A, principle 4.

requirements is governed by rules that make access equitable for all. This has in fact been acknowledged by the Planning Commission which has already suggested limitations on the quantity of water landowners can appropriate and to consider the rest as a community resource.[44]

In other words, delinking water and land should be done to foster equity and sustainability rather than to foster efficiency. This requires basing the delinking process on a series of broad principles that guide its implementation. Both the principles of common heritage and public trust provide appropriate starting points to rethink the legal regime in ways that will ensure that it caters in priority to social and environmental aims.

3. Principles for water allocation

The introduction of the principle of common heritage of humankind provides the general structure of a new legal regime. It needs to be supplemented by principles for allocating water given the overall limited supply in a number of locations. As noted in chapter 3, water policies tend to provide a prioritization scheme for water use. These are sometimes inappropriate where the government is allowed to arbitrarily modify the prioritization and generally needs to be supplemented by additional elements.

Firstly, while water policies make a distinction between drinking water, agriculture, and industry, they do not provide the means to distinguish water uses related to the fulfilment of the human right to water, food and health, livelihood activities, and commercial activities. The category 'drinking water' needs to be understood in a much broader sense that includes not only water needed for drinking but also water uses that are indispensable to ward off water-related diseases as well as water uses necessary to prepare food and in rural contexts to produce food as well. Access to this water must take into account the deep inequality that makes women and children bear a disproportionate burden of insufficient access to water. Water allocation must also take into account that basic water uses should always be met from local sources to the extent that this is possible. If other uses cannot be met locally, these must then be either externally sourced or incentives must be given, for instance, to undertake other economic activities. In other words, allocation principles should also be based on proximity.

Secondly, water should be allocated in such a way that it is local development that takes precedence. This is true from the local to the national level. The fulfilment of the human right to water in a given locality should not be trumped by the needs for water of a big city. Additionally, the increasing bias against rural areas must be addressed. Thus, policy is today based on the premise that water must be reallocated from low-value agricultural use to high-value urban and

[44] Planning Commission, Mid-Term Appraisal of the Tenth Five Year Plan (2005) 229.

industrial use.[45] While this is a convenient way to reallocate water from irrigation as the highest user of water to other sectors it fails to take into account that most of the water used in rural areas is in fact indirectly or directly devoted to urban areas. Thus, an increasing percentage of the crops grown under irrigation in rural areas are consumed in urban areas. From this standpoint the imbalance between rural and urban areas is much less obvious than suggested in existing policy proposals.

The priority put on local development also implies that the livelihood needs of a specific district should not be jeopardized by water intensive industrial activities. In other words, water-consumptive activities of national relevance which are not a priority in terms of social or environmental policies at the local level should be given second priority. In many cases, this is only a question of prioritization because there is enough water to meet human rights commitments as well as to undertake economic development activities. This prioritization ensures, however, that water scarce regions are not targeted for water consumptive schemes that may come at the cost of social and environmental needs. Such prioritization does lead in certain cases to conflicts that will need to be adjudicated according to the principles that make up the basic structure of water law.

Thirdly, the framework for water use is the need for solidarity from the local to the national levels. There should thus be no curtailment of someone's right to contribute to the fulfilment of someone else's right. The displacement of oustees by a dam that is premised on the fulfilment of the fundamental right to water of other citizens of the same country constitutes one example of the kind of situation that can arise in this context. Solidarity does not allow such trade-offs and imposes a broader view that looks at development in a way which includes each and everyone, regardless of their social or economic position. Policies or decisions that justify the violation of someone's fundamental right because it contributes to the realization of someone else's right are not acceptable. This is all the more inadmissible in the situation where it is the poorer and economically and socially weaker citizens that have to give up house, livelihood, and land for the benefit of largely better-off urban dwellers.

C. Broader Conception of the Human Right to Water

As highlighted in chapters 2 and 5, the human right to water is a reality of the international and national legal orders. Not only is there no scope for denying the existence of a right so important that its denial would put in question the whole framework of human rights but in fact, there are good grounds to argue that it is part of the peremptory norms of international law. As repeatedly discussed in the

[45] World Bank, India—Water Resources Management Sector Review—Urban Water Supply and Sanitation Report (Report No. 18321, 1998) 8.

literature, human rights have often been proposed as candidates for *ius cogens* status.[46] Indeed, one of the identification mechanisms for peremptory status is that the norm safeguards interests that transcend those of individual states, have a moral connotation and their breach is so morally deplorable as to be considered absolutely unacceptable by the international community as a whole.[47] If some rights like the right to life are often deemed to have the requisite support to be considered *ius cogens*, there is little doubt that the human right to water also enjoys this status.[48]

There is an increasing consensus among all concerned actors that the human right to water exists and must be realized for all.[49] Yet, the multinational companies that now support the recognition of the human right to water do so out of the understanding that this provides a new business opportunity as this will increase the number of their customers in a context where part of the costs of new infrastructure will likely be borne by the public sector.[50] There are thus different conceptions of the scope of the right. These different views make a significant difference to its realization.

1. From drinking water to livelihood water

One of the key questions that arise today in the context of the realization of the human right to water is that of its actual content. This can be divided into two components. Firstly, there is the question of the types of water uses that are covered by the human right. Secondly, one of the points that has attracted most attention is that of the quantity of water necessary for the right to be deemed realized.

Existing clauses concerning the human right to water tend to focus on issues of access to drinking water. As such, the inclusion of drinking water in the human right to water seems beyond question. Most definitions of the right also include a number of domestic uses, and in particular some health-related uses of water. General Comment 15 is representative of a definition that specifically seeks to limit the scope of the human right to water to a relatively narrow set of water uses. This was done on purpose by the Committee which felt that other aspects of the right would be better dealt with under other Covenant rights.[51]

[46] A Bianchi, 'Human Rights and the Magic of *Jus Cogens*' (2008) 19 *European J Intl L* 491.
[47] A Orakhelashvili, *Peremptory Norms in International Law* (Oxford: Oxford University Press, 2006) 50.
[48] ibid 60 for a specific discussion of the case of economic and social rights.
[49] C Dubreuil, Synthesis on the Right to Water—Fourth World Water Forum, Mexico (Marseille: World Water Council, 2006).
[50] cf AK Singh, *Privatization of Rivers in India* (Mumbai: Vikas Adhyayan Kendra, 2004) 40.
[51] E Riedel, 'The Human Right to Water and General Comment No. 15 of the CESCR' in E Riedel & P Rothen (eds), *The Human Right to Water* (Berlin: Berliner Wissenschafts-Verlag, 2006) 19, 33.

The limitation of the right's scope to drinking water, even in its broader understanding as domestic water, is problematic because this does not in itself ensure a life of dignity. A broader framework thus needs to be adopted. Indeed, for a majority of the world's population, water is not only a question of mere survival in relation to drinking water but also intrinsically linked to their livelihoods. Water linked to livelihood use and subsistence food production is thus part of the water uses that must be considered. These include water required for growing food crops for one's own use, growing food crops for supply to local markets or the water required as part of a livelihood such as pottery. Water as a livelihood is often more important for people living in rural areas but since the rural population constitutes 57 per cent of the population of developing countries and 71 per cent of India's population, their interests cannot be brushed aside.[52] Additionally, since water is intrinsically related to the production of food, urban dwellers also have a stake in this since their food is produced by their rural counterparts.

In broader terms, the issue that arises is the question of the water necessary to meet the needs of local social and economic development.[53] This is particularly significant in the context of irrigation. Indeed, irrigation water provides both the basis for the water needs of subsistence crops and those of water intensive cash crops. The focus on livelihood does not imply that all irrigation water is covered by the human right to water. Indeed, a distinction must be maintained between water used for irrigation of cash crops that are mainly to be exported and food crops that are to be consumed mostly locally or nationally. The latter fall under the scope of the right to water.[54] The addition of livelihood uses to the scope of the human right to water broadens its scope significantly. This is necessary because restricting the scope of the right to drinking or domestic water does not lead to acceptable results. This is true in terms of the links between the realization of the human right to water and other rights such as the human right to food as well as in terms of the limitations imposed in practice on the quantity of water deemed sufficient to realize the right that is recognized.

The question of quantity of water necessary to realize the right has in fact become one of the key elements of the definition of a human right to water. This is not the best indicator because a quantitative assessment of the realization of a human right cannot provide a comprehensive picture of its implementation. The focus on specific quantities of water is nevertheless useful as one of the indicators of the realization of the right. At the international level, the figure that is often touted as a minimum threshold is 20 lpcd.[55] This does not provide an appropriate

[52] United Nations Development Programme, *Human Development Report 2007/2008* (New York: UNDP, 2007) 243.

[53] eg Petrella (n 32 above) 93.

[54] cf Fundamental Principles for a Framework Convention on the Right to Water (Geneva: Green Cross International, 2005) art 1.

[55] United Nations Development Programme, *Human Development Report 2006—Beyond Scarcity: Power, Poverty and the Global Water Crisis* (New York: UNDP, 2006) 3, 5.

benchmark because the human right to water cannot be deemed to be realized at this level. While our actual survival needs may be of only around 5 litres per day, this does not ensure long-term survival which includes food (and thus the need to cook food with water) as well as various health-related needs from bathing and ablutions to washing clothes. In any case, the point of a human right is not to guarantee mere existence and long-term survival but a life of dignity.[56] In South Africa, the minimum amount of 25 lpcd provided for a household of 8 people regardless of the actual number of residents has been declared 'unreasonable' and raised to 50 litres by the High Court.[57] In India, the prevailing norm for rural areas has been a minimum of 40 lpcd and, with variation, 70 lpcd for urban areas. On the whole, a benchmark is useful in assessing progress towards the realization of the right. Yet, where the benchmark is woefully too low as in the South African case, it ends up serving little purpose in distinguishing realization from violation of the right. In other words, as long as quantities suggested in policy documents are aligned with basic survival needs, assessing the realization of the right from a quantitative perspective will fail because the premise is unacceptable in social and equity terms.

2. Disconnections

The provision of drinking water supply raises the possibility that it may be disconnected. This is particularly important because drinking water is a human right. Disconnection of water services in India has been theoretically possible according to certain laws for a number of years. Yet, even if this was sometimes carried out, this took place in a context where the government was providing both private connections and public standposts. The disconnection would thus have forced affected people to rely on existing public water sources rather than be without access to water altogether.

The question of disconnections has become much more important in the context of water sector reforms. This is due to two main factors. Firstly, the focus on turning water into an economic good means that disconnecting people who do not pay is seen as a way to bring in more revenue and thus financial sustainability of operations. Secondly, water sector reforms are increasingly leading to a situation where the government will not be investing, and in fact in some cases withdrawing, existing infrastructure providing common access to water.[58] This is significant because the implication is that disconnections in the future will have much more severe impacts for affected individuals than would have been the case

[56] J Donnelly, *Universal Human Rights in Theory and Practice* (Ithaca: Cornell University Press, 2nd ed 2003) 14.

[57] *Lindiwe Mazibuko v City of Johannesburg* Case No. 06/13865 (High Court of South Africa, Witwatersrand Local Division, 2008) paras 181, 183.

[58] Concerning the phasing out of handpumps under the Jawaharlal Nehru National Urban Renewal Mission, ch 5.A.2, p 146.

earlier. Under water sector reforms, disconnections are envisaged as a legitimate tool to ensure the implementation of reforms.

In the context of the human right to water, disconnections are significant. In a relatively weak formulation of the right as adopted in the General Comment, it is only 'arbitrary' disconnections that constitute a violation of the right to water.[59] There is thus no bar on disconnections per se. This position was confirmed in a more recent report of the High Commissioner for Human Rights that focuses on the process and safeguards in cases of disconnection rather than on the question of disconnections per se.[60] Even though this is the logical conclusion of a system based on water seen as an economic good, it is problematic from a human right point of view. This is indirectly recognized in the report just mentioned, which makes a distinction between 'full disconnections' that deprive individuals of a 'minimum essential level of water and basic sanitation' and other disconnections.[61] The distinction between disconnection and full disconnection is not explained and is an unhappy one since individuals facing disconnections are unlikely to see any difference between the two. Difficulties caused by disconnections have been recognized in various countries even where the human right to water is not specifically mentioned as a human right. In France, judges have opposed disconnections on the ground that they would be disproportionate measures that deprive users of an element that is essential to life.[62] In Belgium, in a challenge to the adoption of a system in the Brussels region that imposes on building owners co-responsibility for the payment of water charges where tenants default to ensure uninterrupted water supply, the judge opined that this measure was a reasonable social policy tool in view of the importance of access to water.[63] In the Flemish region, users have a right to an uninterrupted minimal service and disconnections can only be effected in case of immediate danger or obvious fraud.[64]

In addition to these examples, the situation of England and Wales and South Africa stand out since they confirm that with or without a human right framework disconnections are unacceptable. In the case of the former, the 1989 privatization directly paved the way for disconnections which were not outlawed as part of the

[59] Committee on Economic, Social and Cultural Rights, General Comment 15: The Right to Water (Articles 11 and 12 of the International Covenant on Economic, Social and Cultural Rights), UN Doc. E/C.12/2002/11 (2003) s 44.

[60] United Nations High Commissioner for Human Rights, Report on the Scope and Content of the Relevant Human Rights Obligations Related to Equitable Access to Safe Drinking Water and Sanitation under International Human Rights Instruments, UN Doc. A/HRC/6/3 (2007).

[61] ibid 24.

[62] *Monsieur François X... Union fédérale des consommateurs d'Avignon c/ Société avignonnaise des eaux* N° 1492/95 (Tribunal de grande instance d'Avignon, Ordonnance de référé, 1995) and *Compagnie de services et d'environnement c/ Usagers* N° 9800223 (Tribunal de grande instance de Privas, Ordonnance de référé, 1998).

[63] *Arrêt N° 9/96 concernant le recours en annulation de l'Article 3 de l'Ordonnance de la Région de Bruxelles-Capitale du 8 septembre 1994 réglementant la fourniture d'eau alimentaire distribuée par réseau en Région bruxelloise*, introduit par l'ASBL Syndicat national des propriétaires et autres (1996).

[64] Communauté Flamande, Décret réglant le droit à la fourniture minimale d'électricité, de gaz et d'eau 1996.

regulatory framework put in place at that point. In fact, the Water Industry Act 1991 specifically provided for the possibility to disconnect a service pipe to an occupier who had failed to pay charges due to the operator, as soon as seven days had elapsed after the serving of a notice.[65] In the context of water bills increasing by 67 per cent between 1989–90 and 1994–5, 12,500 households were disconnected in 1994.[66] This issue came up for adjudication when individuals questioned the legality of pre-payment water devices, also called 'budget payment units'.[67] The Court found that the automatic closure of the valve did cut off supply and that they were unjustifiable under the statutory provisions. A year later, Parliament adopted amendments to the statutory framework that specifically prohibits disconnection for non-payment of water charges to any dwelling which is someone's main home.[68]

In South Africa, the introduction of water sector reforms based on the notion of water as an economic good also opened the door for disconnections. Thus, in Phiri, one of the townships forming Soweto in Johannesburg, the implementation of the free water policy was, for instance, tied to the introduction of pre-payment meters. These measures were challenged in court and the judge made a number of significant observations. He first asserted that the government has an obligation to provide basic water to the poor. He then found that pre-payment meters with automatic shut off mechanisms are unconstitutional and unlawful.[69] Additionally he found that forcing residents to restrict their water usage for such fundamental needs as sanitation amounted to denying them the right to health and to lead a dignified lifestyle.

The above examples confirm that disconnections are not justifiable whether the human right to water is formally recognized or implied. The position taken by UN human rights organs is inappropriate and must be reviewed to reflect the fundamental nature of water.

3. Free water

The question of whether the realization of the human right to water implies the provision of free water has been increasingly discussed in recent years. This is linked to two divergent trends. On the one hand, as part of the formalization of the human right to water in various contexts, the question of related obligations has taken up more importance. On the other hand, the introduction of an economic conception of water has led to the call for full cost recovery, including for drinking water.

The question of free water raises preliminary issues concerning the extent to which human rights must be provided free. Despite earlier claims that there is a

[65] Water Industry Act 1991 (c. 56) s 61.

[66] C Ward, *Reflected in Water—A Crisis of Social Responsibility* (London: Cassell, 1996).

[67] *R v Director General of Water Services* [1999] Env LR 114 (Queen's Bench Division (Crown Office List), 1998).

[68] Water Industry Act 1991 (modified by Water Industry Act 1999) s 61(1A).

[69] *Lindiwe Mazibuko v City of Johannesburg* Case No. 06/13865 (High Court of South Africa, Witwatersrand Local Division, 2008) para 183.

difference between first and second generation rights, the realization of all rights requires state intervention and the allocation of public resources. In fact, all rights have a cost even the ones that, like the right to property, are mostly associated with the state's duty to refrain from interfering with freedoms.[70] Yet, not all states treat all rights in the same way. Thus, while the Covenant specifically provides that primary education must be provided free,[71] no similar consensus exists with regard to food and health which are much more directly related to survival. The right to food is noteworthy because it is usually not provided free but contexts of poverty have, for instance, led the Supreme Court to direct that free mid-day meals should be provided in all schools.[72] The right to health provides yet a different story. In India, the existing public sector health care dispenses healthcare in a subsidized manner or freely to the poor.[73] The main problem is that the public sector healthcare is undersized. There is, for instance, inadequate availability of trained healthcare professionals for rural areas.[74] This can be compared to the situation of the United Kingdom where a massive free public health system is provided through taxation.

A case may be made that the core content of all human rights should be provided free. This is not pursued here because in the context of water a different equation arises. Indeed, if nearly all civil and political rights are provided free there are good reasons to believe that the core content of a right much more fundamental in terms of survival like the right to water should be provided free. The question then boils down to the ways in which a right can be provided free. In most cases, the argument against providing free water is that the state cannot afford such expenditure. The real issue, however, is not whether the state has resources but how the state decides to allocate available resources according to identified policy priorities. The realization of human rights is without the slightest doubt of greater importance than the defence budget and in many countries the reallocation of a small share of the defence budget would go a long way towards the realization of many human rights.

Turning to water, the provision of free water should be an uncontroversial proposition.[75] The successful implementation of free health policies confirms that there are no conceptual problems. The reluctance of the General Comment to propose a free water policy as part of the core aspects of the right to water thus

[70] eg S Holmes and CR Sunstein, *The Cost of Rights—Why Liberty Depends on Taxes* (New York: WW Norton, 1999).

[71] International Covenant on Economic, Social and Cultural Rights, New York, 16 December 1966, 993 UNTS 3 (1976), art 13(2).

[72] J Drèze, 'Democracy and the Right to Food', in P Alston & M Robinson (eds), *Human Rights and Development—Towards Mutual Reinforcement* (Oxford: Oxford University Press, 2005) 45.

[73] Concerning Kerala, K Praveenlal et al., 'Healthcare and User Charges' (2005) 40/7 *Economic & Political Weekly* 615.

[74] Planning Commission—Government of India, *Eleventh Five Year Plan 2007–12—Volume II—Social Sector* (New Delhi: Oxford University Press, 2008) 64.

[75] However H Smets, 'Le droit à l'eau, un droit pour tous en Europe' (2007) 37/2-3 *Environmental Policy & L* 223, 223 argues that the right to water is not a right to free water for all.

cannot be ascribed to an impossibility to implement free water entitlements. Rather, it must be understood as a consequence of the focus on water as an economic good. From this perspective, in terms of economic efficiency, free water is seen as a bad proposition that may lead to additional water wastages.[76] Another argument that is put forward against free water is that states cannot afford the cost.[77] Yet, the provision of free water as part of the right to water is not unknown in UN human rights documents. Thus, the draft guiding principles on extreme poverty and human rights specifically provide that drinking water must be provided free of cost.[78]

Additionally, there already exist countries where water is free. In Ireland, for instance, water services are entirely free for household use.[79] This is limited to domestic requirements and does not extend to agriculture or commercial activities.[80] If Ireland has one of the most progressive stances with regard to the realization of the human right to water, it is South Africa and its restrictive free water policy that has attracted most attention in recent years. The South African free water policy introduced in 2001 committed the government to provide 25 litres per person or 6 kilolitres per household per month within 200 metres of a household for at least 358 days a year.[81] The South African policy has been much debated because it is alternatively seen as progressive and restrictive. Divergent analyses are due to the fact that post-apartheid water law is at the same time based on the principle of cost recovery and on the realization of the human right to water. Indeed, the Water Services Act proceeds from a logic that is largely different from the fundamental right to water recognized in the Constitution.[82] More broadly, water policy has been marred by the tension between the focus on poverty alleviation and the neo-liberal economic policy agenda of the government highlighted, for instance, in the Growth, Employment and Redistribution Strategy.[83] While the Water Act 1998 was a major step forward in its recognition that water is a public trust and is not linked to property rights in land, the Water

[76] SMA Salman & S McInerney-Lankford, *The Human Right to Water—Legal and Policy Dimensions* (Washington, DC: World Bank, 2004) 70 argue that '[f]ree water is an invitation for misuse and abuse'.

[77] cf SC McCaffrey, 'The Human Right to Water' in E Brown Weiss, L Boisson de Chazournes & N Bernasconi-Osterwalder (eds), *Fresh Water and International Economic Law* (Oxford: Oxford University Press, 2005) 93.

[78] Draft Guiding Principles 'Extreme Poverty and Human Rights: The Rights of the Poor', Resolution 2006/9, Implementation of Existing Human Rights Norms and Standards in the Context of the Fight Against Extreme Poverty, in Report of the Sub-Commission on the Promotion and Protection of Human Rights on its Fifty-Eighth Session, UN Doc. A/HRC/2/2-A/HRC/Sub.1/58/36 (2006) principle 29.

[79] Ireland, Water Services Act 2007, s 105(1).

[80] ibid s 105(12).

[81] South Africa, Regulations Relating to Compulsory National Standards and Measures to Conserve Water 2001, s 3.

[82] South Africa, Constitution, s 27(1)(b).

[83] Department of Finance—Republic of South Africa, Growth, Employment and Redistribution—A Macroeconomic Strategy (1996).

Services Act 1997 did not include a free water policy at the outset. Free water was only eventually adopted in 2001 after a serious cholera epidemic which was at least partly caused by the unavailability of clean water after the fitting of pre-paid water meters.[84]

The implementation of the free water policy has also been marred by controversy. Municipalities can choose different options to fulfil their mandate. They can decide to provide the free amount to everyone, only to the poor or decide to provide it to everyone but make people who consume more than the free amount pay not only for their additional consumption but also for the water that is in principle provided free.[85] This links up with the quantity benchmark originally adopted. 25 litres per day is in most contexts insufficient for a dignified life. This is especially the case in a city environment where water has to be used for such basic uses as latrines and washing clothes. Additionally, the South African policy unfortunately failed to really provide 25 litres to each individual. The possibility for municipalities to provide an amount per household means that any household that has more than eight individuals would get less than 25 litres per person per day. The problem associated with this quantity was known and acknowledged at the time the post-apartheid legislative framework was being conceived. Thus, the ANC's Reconstruction and Development Programme recognized that the provision of 20 to 30 lpcd within 200 metres of the household was a short-term goal which was to be expanded in the medium-term to an on-site supply of 50 to 60 litres.[86] As noted above, in the *Mazibuko* case, the High Court ordered that 50 lpcd be provided as free water.[87]

The Irish and South African examples confirm that there are neither conceptual barriers to providing free water for domestic use nor any economic reasons that would explain why the state cannot invest its resources in water infrastructure. The situation of each country needs to be analysed separately and it is likely the Irish model cannot be directly implemented in India. Yet, in a context where urban and rural areas are in any case conceived as separate realms, the Irish model could be applied to urban areas where monitoring is much easier to achieve. In fact, the JNNURM and UIDSSMT reforms that put emphasis on water supply may provide the physical backbone of new delivery mechanisms. In rural areas, the provision of free water has not been accused of unfairly subsidizing commercial use of water since there is little scope for this to happen in practice. Free water has in any case been provided as a matter of policy in rural areas in India and there is thus no need to demonstrate its feasibility.

[84] JA Smith and JM Green, 'Free Basic Water in Msunduzi, KwaZulu-Natal: Is it Making a Difference to the Lives of Low-income Households?' (2005) 7 *Water Policy* 443.

[85] ibid 446.

[86] African National Congress, The Reconstruction and Development Programme—A Policy Framework 1994 ss 2(6)(6–7).

[87] *Lindiwe Mazibuko v City of Johannesburg* Case No. 06/13865 (High Court of South Africa, Witwatersrand Local Division, 2008) para 183.

4. Links with other human rights

The human right to water is linked to a number of other human rights and its realization is a precondition for the realization of a number of rights. Firstly, the human right to water is indissociably linked to the right to life. This is borne out of the decisions of the Supreme Court that first derived a right to water from the right to life.[88] The link has also been specifically made at the international level.[89]

Secondly, the right is linked to several socio-economic rights such as the rights to health and food. Links arise from the fact that they constitute some of the most basic elements that ensure a life of dignity and some of the utmost priorities for equitable socio-economic development. The links also surface insofar as the realization of each of the three rights depends in large part upon the realization of the other two. This confirms not only the impossibility to recognize one and deny the others but also the need to take an integrated view of their realization. This raises important issues for the realization of the human right to water. Indeed, if the scope is understood in the narrow conception underlined earlier, its scope may not include the use of water for subsistence food production. Since water is an essential input in agriculture, there is a direct link between the production of sufficient food crops and water availability. The argument against this is that countries experiencing physical water scarcity cannot implement such a right.[90] This is, however, not an appropriate starting point of enquiry. Few countries are physically water-scarce at this juncture. Even though more may become so in the future, this does not provide an answer from a human rights point of view. Any country that is unable to provide water to grow food crops today has to purchase food to ensure the survival of its population and thus contribute to the realization of the right to food. The resources diverted to purchasing food could be redirected in part to foster water availability for growing basic food crops. In any case, apart from the most extreme cases, water scarcity is not absolute. The main issue is the allocation of water uses between different sectors. Since irrigation for cash crops such as cotton or sugarcane is by far the main user of water, the issue usually boils down to one of allocation or of addressing the causes of social or economic scarcity. In a minority of cases, countries may need to ensure the joint realization of the rights to food and water in a different way, something that falls within the margin of appreciation that human rights courts grant to national governments.[91]

[88] *Subhash Kumar v State of Bihar* AIR 1991 SC 420 (Supreme Court of India, 1991).

[89] eg Draft Guiding Principles 'Extreme Poverty and Human Rights: The Rights of the Poor', Resolution 2006/9, Implementation of Existing Human Rights Norms and Standards in the Context of the Fight Against Extreme Poverty, in Report of the Sub-Commission on the Promotion and Protection of Human Rights on its Fifty-Eighth Session, UN Doc. A/HRC/2/2-A/HRC/Sub.1/58/36 (2006) principle 30.

[90] T Kiefer & C Brolmann, 'Beyond State Sovereignty: The Human Right to Water' (2005) 5/3 *Non-State Actors & Intl L* 183.

[91] eg *Fadeyeva v Russia* Application No 55723/00 (European Court of Human Rights, 2005).

Thirdly, the realization of the human right to water is also related to the right to a clean environment. There is a direct link between environmental protection and the realization of the human right to water because its realization depends on sufficient water flows and the availability of water of an acceptable quality. Environmental concerns arise at different levels concerning both surface and groundwater. A specific example of the link between the right to a clean environment and water, concerns water flows in rivers since insufficient flows are one of the causes of reduced water quality and all ensuing health problems. This is well illustrated in the case of Delhi that sees an increasingly reduced Yamuna river flow through the city. This affects water supply for Delhi residents and dramatically reduces the ability of the river to support the pollution load that it receives in the form of sewage. In this context, the right to a clean environment and water are partly indissociable. This is often recognized in law and policy since a number of water-related issues are addressed in the context of environmental laws and institutions set up to implement them.

Fourthly, several procedural rights, such as access to information and participation are also closely involved in the realization of the human right to water. Access to information in an environmental or water context is still a new topic in India. For the time being, the only specific framework that applies is the Right to Information Act 2005 which upturns the previous system where information was not accessible as a matter of right. This constitutes a major step forward in ensuring access but remains conceptually limited because it is conceived around a system where citizens seek information from the government rather than a system where the government must provide information without being prompted. A more specific framework is the Aarhus Convention, which specifically concerns access to environmental information.[92] The scope of information covered under the Convention is quite broad, encompassing a non-exhaustive list of environmental elements, including water, and factors likely to affect the environment. Access to environmental information is based on a presumption in favour of disclosure subject to explicit—and restricted—exceptions such as confidentiality. The Convention makes it clear that the information disclosed or disseminated must be relevant, adequate and understandable, and must be made available in a transparent way and effectively accessible. Emphasis on access to information is made much more important by the fact that proposed reforms may restrict existing rights of access. Thus, the Uttar Pradesh Water Management and Regulatory Commission Bill 2008 specifically proposes that information in possession of the Commission must be kept confidential and that all information obtained by the Commission must be treated as classified.[93] This will dramatically restrict existing rights to access information, something which is neither desirable nor acceptable.

[92] Convention on Access to Information, Public Participation in Decision-Making and Access to Justice in Environmental Matters, Aarhus, 25 June 1998, UN Doc. ECE/CEP/43.

[93] Uttar Pradesh Water Management and Regulatory Commission Bill 2008, s 17.

Participation also plays a direct role. The importance of participation in this context is that it is in principle the same concept found in the context of water sector reforms. In reality, participation in a human right context refers to a different notion, which should not be called by the same name as it creates confusion. It refers here to a comprehensive set of measures that give everyone—and not a selected group of people—the same rights. It also refers to rights that people have and there is thus no need to create new rights. Existing participatory rights may need better implementation and enforcement but their existence for everyone is beyond doubt. An important dimension of participation is reflected in the principle of prior informed consent, which finds its source in international environmental agreements such as the Biodiversity Convention.[94] It has been accepted and become a part of domestic law with the adoption of the Biodiversity Act 2002.[95] Yet, it has only been applied in the context of inter-state relations whereas the principle has a much broader meaning which encompasses the consent of the concerned people. This is progressively being accepted at the international level, as illustrated by the United Nations Declaration on the Rights of Indigenous Peoples.[96]

D. Rethinking Water Sector Principles

Water sector reforms are based on a set of principles that should in theory be acceptable to all as a general proposition. Yet, the emphasis of water sector reforms on the economic nature of water gives the policy framework adopted to realize these reforms a specific direction. The focus on efficiency as a way to bring about equity or the use of the environment as a basis to bring about largely economic reforms do not constitute the only way to address these issues. This section consequently reassesses some of the basic principles of ongoing water sector reforms to foster a broader debate on the policy choices that countries make towards the introduction of water law reforms for the twenty-first century.

1. Equity

The focus on efficiency as a way to achieve equity in ongoing water sector reforms has led to conceiving equity narrowly.[97] This is unfortunate because equity is of central importance in relation to water in its multiple dimensions. Equity is useful as a construct because it pervades the legal system. As developed by judges it provides an instrument to ensure that the strict application of the law is not unfair

[94] Convention on Biological Diversity, Rio de Janeiro, 5 June 1992, Doc. UNEP/CBD/94/1.

[95] Biological Diversity Act 2002, c 3.

[96] United Nations Declaration on the Rights of Indigenous Peoples, General Assembly Resolution 61/295, 13 September 2007, UN Doc. A/RES/61/295, art 32(2).

[97] cf Ingram, Whiteley & Perry (n 1 above) 23 stating that '[e]conomic returns not community value of water or other equity issues drive consideration of efficiency'.

in a specific case. In recent decades the understanding of equity has evolved.[98] It goes beyond providing a way to avoid unjust results in the application of formally just rules and offers the basis for the adoption of rules that take into account systemic inequities. Equity thus moves from a focus on formal equality as the basis for fair results to a focus on substantive equality as the standard against which rules must be judged. In other words, equity now provides the basis for adopting measures that discriminate in favour of the weaker, marginalized sections of society on a systematic basis to ensure that the application of the rules adopted lead to a result which is just and fair according to the broader understanding of justice and fairness that underlies the legal system. In India, reservation policies constitute one of the instruments used to address equity concerns.[99] This is embodied in the Constitution that authorizes the preferential treatment of socially and educationally backward classes, schedules castes and scheduled tribes.[100]

In the context of water, equity provides, for instance, a way to link poverty, discrimination in access to water and the need for preferential measures to ensure that the poorest and weakest are not further deprived because of their socio-economic conditions.

Firstly, equity can be used to address discrimination among individuals. In the case of water, different groups of people need to be singled out for special measures. These include women who bear most of the burden of fetching water, children and in particular the girl child, people subject to discrimination in access to water because of their caste or religion, and people in extreme poverty. The relevance of gender as an independent factor in assessing the impacts of water sector reforms has, for instance, been acknowledged in South Africa. In the *Mazibuko* case, the High Court took specific notice of the special burden that prepayment meters impose on women since the responsibility of fetching water falls on them.[101] In other words, access to water raises substantive concerns with regard to the right to non-discrimination. There is thus a clear link between equity and human rights that crystallizes most clearly in the context of the fundamental right of non-discrimination, a right clearly articulated at the national and international levels.[102]

Equity concerns also arise with regard to people situated in different locations. This is, for instance, illustrated by the case of big dams. On the one hand, people ousted by dams lose their home, livelihood, access to resources and access to water. In an ideal world, oustees would be the first to benefit from the project that displaces them and thus at least regain the standard of living they were enjoying

[98] TM Franck, *Fairness in International Law and Institutions* (Oxford: Clarendon, 1995).

[99] eg M Galanter, *Competing Equalities—Law and the Backward Classes in India* (Berkeley: UC Press, 1984).

[100] Constitution, art 15.

[101] *Lindiwe Mazibuko v City of Johannesburg* Case No. 06/13865 (High Court of South Africa, Witwatersrand Local Division, 2008) para 159.

[102] eg Indian Constitution, art 14 and International Covenant on Civil and Political Rights, 16 December 1966, 999 UNTS 171, art 26.

before. In practice, this has often proved not to be the case.[103] On the other hand, dams increasingly often include a drinking water component, as in the case of the Sardar Sarovar dam in Gujarat.[104] Part of the water diverted by the dam is thus used to meet the drinking water needs of rural or urban residents in other areas. The dichotomy between the curtailment of some people's human rights that indirectly contributes to realizing the human rights of some other people can either be seen as a conflict of rights or as an equity issue. In the Sardar Sarovar case, Justice Kirpal emphasized that the dam contributes to fulfilling the unrealized fundamental right to water of the beneficiaries in the command area.[105] In this view, the sacrifice of the oustees is seen as a necessary evil for the progress of the nation. This is doubly problematic because it does not put on the same level the rights of the oustees and the beneficiaries of the new drinking water supply. Additionally, it fails to consider the human right to water of the oustees. There is dark irony in this since oustees that live near a river tend to depend on it for most of their water needs in direct and indirect ways. Since any fundamental right is the right of each and every individual, there can be no prioritization in favour of certain groups of people. Indeed, from the point of view of the human right to water, a scheme like the Sardar Sarovar dam is only justifiable if it firstly contributes to the realization of the human right to water of the oustees who are often among the poorest, and subsidiarily also contributes to improve the situation of other people who are not affected by the dam.

Secondly, equity provides a useful starting point to discuss the acceptability of measures that discriminate among groups of people according to their location or socio-economic profile. In the case of water, one of the issues that stands out is the distinction that is made in policy terms between urban and rural areas. The existing policy framework provides that better-off parts of the country—cities—deserve a higher per capita allocation of domestic water. The rationale for providing urban dwellers with more water per capita than villages and for providing residents of bigger cities with more water than smaller cities is linked to the availability of sewage systems. Urban areas have also been given a disproportionate proportion of water supply investments for quite some time.[106] This does not seem to have changed as the Planning Commission is still calling on states in the context of the eleventh plan to give 'topmost priority' to the drinking water needs of urban areas.[107]

[103] eg Tata Institute of Social Sciences, Performance and Development Effectiveness of the Sardar Sarovar Project (Mumbai: TISS, 2008).

[104] eg I Hirway & S Goswami, 'Functioning of the Drinking Water Component of the Narmada Pipeline Project in Gujarat' (2008) 43/9 *Economic & Political Weekly* 51.

[105] *Narmada Bachao Andolan v Union of India* AIR 2000 SC 3751 (Supreme Court of India, 2000).

[106] World Bank, India—Water Resources Management Sector Review—Initiating and Sustaining Water Sector Reforms (Report No. 18356-IN, 1998) 32 indicates that urban areas received 40% of drinking water investments at a time when urban residents were just over a quarter of the overall population.

[107] Planning Commission—Government of India, *Eleventh Five Year Plan 2007–12—Volume II—Social Sector* (New Delhi: Oxford University Press, 2008) 117.

The absence of effective water-based sanitation as a ground for lower water allocation is an inappropriate response to a serious problem and raises anti-discrimination concerns. Indeed, since water-based sanitation is an integral part of a holistic water system, a lower allocation in policy indicates that sanitation will not be provided in the short or medium-term, as non-water based alternatives are not yet a solution for the majority of the population. This is also problematic because it rewards wasteful water use patterns among affluent sections of bigger towns. Further, a lesser burden is put on more affluent urban communities who are not asked to make capital costs contributions but only to pay the tariffs imposed.[108] This begs the broader question of the link between water supply and sanitation and the need to conceive both together. In reality, given the abysmally low level of provision of sanitation in India, it is inappropriate to link the two since that would mean putting off for many years the provision of sufficient water to everyone. While the link between water supply and sanitation exists and must be made, the absence of the latter cannot become an excuse for not providing the former.

There are few signs of a change in policy with regard to the differential provision of water to urban and rural areas. Yet, in a sign that this is at least an issue which is and will be considered in the future, the draft water policy for Assam would impose a similar standard for urban and rural areas.[109] At the same time, it is worrying that Uttar Pradesh proposes to also distinguish among rural inhabitants and provide people who can afford an individual house connection 70 lpcd while others will have to contend with 55 lpcd.[110]

Thirdly, equity is also required as a benchmark for water use allocation within and among sectors. In terms of intra-sectoral allocation, difficult questions arise, for instance, concerning the choice of crops that must be grown within the existing water availability. Since the choices that are made have broad-ranging repercussions, decisions must ensure that the result is fair and sustainable. Equity considerations call for the allocation of water first to subsistence crops that feed the local population with a preference for less water intensive crops over more water intensive water crops followed by food crops for distribution outside the local context, with water-intensive cash crops getting the least priority.

Inter-sectoral allocations have been a focus of increasing attention in recent years. There has been a tendency to conceive allocation based on 'competitive' uses. It has, for instance, been suggested that part of the water used in agriculture could be reallocated to drinking water in urban areas. This is an inappropriate way to conceive the issue. The justification for reallocation should, in principle, be the

[108] D Hemson, 'Water for All: From Firm Promises to "New Realism"?' in D Hemson et al. (eds), *Poverty and Water—Explorations of the Reciprocal Relationship* (London: Zed Books, 2008) 13, 28.

[109] Draft State Water Policy of Assam 2007, para 8(4).

[110] Memorandum of Understanding between the State Government of Uttar Pradesh and the Department of Drinking Water Supply, Ministry of Rural Development, Government of India (2007) s 5(iv)(d).

needs of poor urban dwellers who do not have access to sufficient water. Yet, these are the same people whose food needs are also not met or only partially met. Water transfers from one sector to the other must thus not be conceived in terms of an opposition between agriculture and domestic water needs. Rather, equity requires that ways should be found to save the comparatively small amount of water required for the whole population to meet its domestic needs through a combination of measures. This would include transfers from agriculture that could, for instance, take the form of shifts to crops requiring less water or from water-intensive cash crops to less water-intensive food crops. This would also include checking wasteful water use such as superfluous or luxury consumption that may include a number of activities ranging from swimming pools to golf courses.

Fourthly, equity also arises in the context of inter-state allocations. This has been widely debated in the context of some of the most intractable inter-state disputes such as the seemingly never-ending issues raised between Karnataka and Tamil Nadu concerning the Cauvery River.[111] This illustrates one limitation of the current legal framework that does not impose on states an equity framework for discussing issues related to the sharing of waters. The lack of principles that impose specific prioritization, a focus on drinking water and on livelihood leaves the whole system at the mercy of unhelpful politicization.[112] Beyond the eye-catching headlines caused by formal legal disputes, a host of other issues need to find principled solutions. Inter-state water transfers to meet the drinking water needs of a city are a case in point. This is, for instance, illustrated by the case of the Tehri dam in Uttarakhand and water transfers to the city of Delhi. The construction of the dam has seen people displaced facing curtailment of a number of their fundamental rights and environmental damage caused by the dam to ensure better water supply to the inhabitants of the city of Delhi.[113] At the same time, over the course of the past decade, a process of so-called beautification and modernization of Delhi has led to the eviction of thousands of people living in slums and the prohibition of livelihood activities of many others in the name of pollution control.[114] All these oustees who have either moved to the edge of Delhi or in adjoining states are the very people who should have been the first beneficiaries of the additional drinking water provided by the Tehri dam. In fact, it is these urban oustees who provided a moral justification for a transfer of water, which could also have been used for livelihood activities in Uttarakhand. Additionally, it is

[111] RR Iyer, 'Cauvery Award—Some Questions and Answers' (2007) 42/8 *Economic & Political Weekly* 639.

[112] On the controversy surrounding the Hogenekkal drinking water scheme in Tamil Nadu, eg TS Subramanian, 'Troubled Waters' (2008) 25/8 *Frontline*, available at <http://www.flonnet.com/fl2508/stories/20080425250802800.htm>.

[113] eg S Pathak, 'Tehri Dam: Submersion of a Town, Not of an Idea' (2005) 40/33 *Economic & Political Weekly* 3637.

[114] U Ramanathan, 'Demolition Drive' (2005) 40/27 *Economic & Political Weekly* 2908.

debatable whether the social, environmental and economic costs of transferring water from Tehri to Delhi outweigh the cost that Delhi would incur by, for instance, investing in a comprehensive sewerage system that would drastically curtail groundwater pollution in and around Delhi whose benefits would extend to millions of people, including people living downstream near the Yamuna.

As illustrated in the preceding paragraphs, considerations of equity at the national level abound. These are supplemented by similar concerns at the international level which have crystallized in the general principle of solidarity.[115] This is supplemented by the more specific notion of differential treatment.[116] On the basis of the vastly different situations of states in a system based on formal legality, differential treatment provides a basis for rules that include preferences for weaker countries. This is as relevant in the context of global warming as in the context of water.[117] Differential treatment related to water applies in a variety of situations. These include questions ranging from the sharing of a transboundary watercourse or transboundary aquifer to issues related to the impacts of legal instruments concerning the implementation of water-related development projects and the impacts of trade-related treaties such as GATT or GATS on access to water or sanitation for individuals in developing countries. Differential treatment also applies concerning some of the most central issues of water law such as access to clean drinking water. The discrepancy between developed and developing countries and much more so least developed countries in access to water constitutes one of the indicators of the need for special measures in favour of these countries in water-related legal instruments. While water availability remains to a large extent a national issue given the difficulties associated with transporting water over long distances, the realization of the human right to water is an issue of international relevance.[118]

Differential treatment also has an inter-generational dimension, which brings in another perspective to the concept of reserve. Inter-generational equity imposes, for instance, that water abstraction should be kept within limits that will allow future generations sufficient access to water for their own needs. Groundwater abstraction beyond the annual recharge potential is one issue that can be addressed through inter-generational equity.

[115] RSJ McDonald, 'Solidarity in the Practice and Discourse of Public International Law' (1996) 8 *Pace Intl L Rev* 259.

[116] See generally P Cullet, *Differential Treatment in International Environmental Law* (Aldershot: Ashgate, 2003).

[117] The Brussels Declaration, Third United Nations Conference on the Least Developed Countries, 20 May 2001, UN Doc. A/CONF.191/12, para 5 recognizes, for instance, that achieving the supply of safe drinking water must be undertaken in accordance with common but differentiated responsibilities.

[118] General Comment 15 (n 59 above) ss 30ff.

2. Environment and sustainability

The environment has been at the centre of water sector reforms. Yet, as highlighted earlier, environmental issues are little more than the premise for a series of measures that do not effectively address water conservation and sustainable use. The emphasis on efficient water use notwithstanding, there is little emphasis on ecosystem needs with regard to water because economic efficiency seems to predominate in current policy thinking. This is problematic because a focus on water as an economic good may have the collateral impact of fostering higher water use. Indeed, where private sector actors recover the capital costs of their investments by charging users for the water used, users may be encouraged to use more rather than less water.[119]

In the future, the environment needs to be given a key role in the measures adopted to reform water law. This stems from the fact that human and ecosystems needs cannot be dissociated and because the realization of the human right to water depends in large part on the realization of the human right to a clean environment. This must start with the recognition that a specific percentage of existing water must be set aside because not doing so will jeopardize ecosystems and in turn threaten human water uses. This is justified both on the ground of environmental conservation and on the ground of the sustainability of water flows. It is based on some of the most basic principles of environmental law such as prevention and precaution. The principle of prevention is sufficient to address situations where diversion of water for large-scale irrigation does not leave a river with sufficient flow to perform its ecological functions, as well as any other functions that may be required because of human activities, such as reducing the pollution load caused by humans. The principle of precaution is applicable to situations where the full extent of the potential damage may be difficult to assess in advance. This includes in particular impacts that are not fully understood of activities that may contribute to deteriorating groundwater quality.

The need to keep a part of the water aside changes the perspective on inter-sectoral allocation since it specifically indicates that a percentage of water cannot be allocated for human uses. In this sense it complements the framework provided by the human right to water which essentially provides that the water required to fulfil the right cannot be allocated to any other use. This is in fact what South Africa has done though it does not specifically link this to human rights obligations. Post-apartheid water law provides that there must be an 'ecological' and a 'basic needs' reserve.[120] In a world where increasing physical scarcity is to be expected at least in some regions, the concept of reserve is a necessary addition to a modern water law.

[119] D Takacs, 'Environmental Aspects of Water Sector Reforms', in P Cullet, A Gowlland-Gualtieri, R Madhav & U Ramanathan (eds), *Water Law for the Twenty-first Century: National and International Aspects of Water Law Reforms in India* (Abingdon: Routledge, forthcoming 2009).

[120] South Africa, National Water Act 1998, pt III.

It is not actually new since the existence of the human right to water de facto creates a reserve to meet the human right-related water needs but it gives a new face to the concept and is useful in an environmental context.

The concept of reserve is a starting point for environment regulation of water uses. It needs to be supplemented with more specific instruments based on basic environmental principles. Thus, there is a need for an environmental law basis to decisions on water use, for instance, in the case of irrigation. Allocation of irrigation water must first be based on social needs and subsistence crops get the first priority followed by food crops more generally. Additionally, crops relevant for the local population get priority over export crops. Secondly, each crop must also be selected according to environmental considerations. Where there are different varieties of a given subsistence crop, such as wet or dry paddy or where there is a culturally available choice between two crops whose water intensity is different, millet and paddy for instance, the choice must go to the less water intensive crop.[121] Similarly, where a choice must be made between two crops that have a relatively similar water intensity, the one that requires a lesser application of chemical fertilizers, pesticides, and insecticides must be preferred. In other words, an environmental law perspective to water brings in important synergies between the implementation of environment and water laws.

Indian law still needs to develop such ideas but lineaments of this thinking can be gleaned in some recent laws. This is, for instance, the case of a provision that allows a water association to determine the crops they want to authorize within their water entitlement.[122] This does not yet put an environmental context to the decision but at least recognizes that choices can and must be made. Similarly, Andhra Pradesh links water conservation measures and tree plantation and maintenance.[123]

An environmental law perspective to water regulation also brings into relief the fact that it is not just actual access to water which is a relevant factor in judging the impact of government policies and action. Indeed, water availability is also a function of the environmental policies of the state and of the implementation and enforcement of environmental laws. There is, for instance, a direct link between the availability of groundwater and forest laws since trees play a major role in groundwater recharge.

3. Democratic decentralization

As highlighted in earlier chapters, the notions of participation and decentralization are key concepts in the context of water sector reforms. In fact, user

[121] The advantages of wet paddy over dry paddy have, for instance, been increased by the focus of the Green Revolution on wet varieties. Incentives for better dry paddy varieties could thus help in making the latter more attractive overall. cf LC Yah & C-Y Lim, *Southeast Asia: The Long Road Ahead* (Singapore: World Scientific, 2nd ed. 2004) 65.

[122] Maharashtra Management of Irrigation Systems by Farmers Act 2005, s 24.

[123] Andhra Pradesh Water, Land and Trees Act, 2002.

participation is to a large extent what makes reforms palatable to the public at large. The difficulty is that the understanding of participation and decentralization propounded by water sector reforms only partly coincide with the process of democratic decentralization triggered by the adoption of the 73rd and 74th amendments to the Constitution.

The problems triggered by the setting up of institutions like water user associations is that they do not fall under the jurisdiction of the panchayats. Further, they have membership criteria, which do not correspond to the constitutionally sanctioned standards in terms of gender or minority representation. The lack of a coordinating mechanism between panchayats and user associations provides the basis for shifts in control over water resources away from the panchayat which are not specifically acknowledged.

Additionally, the decentralization instituted under water sector reforms is much weaker than the constitutional scheme. In the case of water user associations, for instance, the focus on management issues leads to a framework where the associations are given certain rights and a number of obligations. It is arguable that the latter outweigh the former whereas a comprehensive process of decentralization as envisaged under the 73rd amendment can only strengthen local institutions and democratic governance generally.

The Maharashtra case illustrates the problems that arise with the setting up of institutions outside of the normal constitutional scheme. Indeed, water user associations, unlike panchayati raj institutions are not permanent. Their existence depends on decisions taken at a higher level and they can thus be amalgamated or divided.[124] This lack of stability is detrimental because it gives higher level institutions immense powers over the local associations that are not counterbalanced by effective accountability frameworks.[125]

If the existing form of decentralization is not the best answer to real concerns, this does not mean that decentralization per se is unwelcome. In fact, it should indeed be one of the priorities of the broad governance agenda in the country in general and is not specific to water. While decentralization still needs to be effectively realized in many areas and many places in the country, the major difference between India and many other countries in the South is that the framework for democratic decentralization exists and is clearly established in the Constitution. In a perfect world, a lot more could be wished for and forms of direct democracy could be added to the existing framework. This can, however, be left for later debates. For the time being, the crux of the issue is to ensure that the 73rd and 74th amendments are effectively implemented in all states. At the constitutional level, water has been addressed and it is abundantly clear that panchayats have significant powers over water. At the state level, the process of implementation of the constitutional provisions remains uneven and further

[124] Maharashtra Management of Irrigation Systems by Farmers Act 2005, s 5.
[125] ch 4.A, pp 112 and 116.

measures need to be taken in a number of cases to make panchayat control over water more real.

The central point that must form the basis for future measures is that there is no need for bodies of the kind that water sector reforms propose. This does not mean that administrative divisions are perfect from the point of view of the conservation and sustainable use of a watershed. The same obtains at the level of river basins since the reorganization of states that occurred after independence was not done on a hydrologic basis. The fact that linguistic factors prevailed over other factors such as hydrology may be bemoaned but in any case, no division could have satisfied all needs. Thus, while neither panchayat, nor block, nor district nor state boundaries correspond to hydrologic boundaries, these are the boundaries that are relevant for all issues, whether they concern political rights such as elections or development issues, such as agriculture, forests or water.

No institutional division will ever be able to provide all the answers that need to be given to the conservation and use of resources. In any case, redrawing the map of panchayats on a hydrologic basis would then lead to disfunctional systems such as where two neighbouring panchayats sharing a watershed are in two different states. The need is thus to ensure that there are coordination mechanisms between institutions at the same level—for instance two panchayats in the case of a watershed shared only by them—and between institutions at different levels—for instance, panchayats and districts in the case of a watershed covering several panchayats in two districts. The need for horizontal and vertical coordination is nothing particular since it is a core element of a federal system.[126] There is thus no need for special institutions to address issues that are already within the mandate of decentralized institutions.

The need to work within the existing institutional system does not imply in any way that the need for an integrated and ecosystem approach is sidelined. On the contrary, effective decentralization that is both equitable and environmentally sustainable must be based on sound environmental principles. This is in principle easy to achieve given that the existing environmental law framework already provides the answers. The difficulty lies in ensuring effective cooperation and implementation between ministries and departments dealing separately with overlapping issues. At the national level, this means, for instance, that there is a need for better cooperation between the Ministry of Environment and Forests, the Ministry of Water Resources, the Ministry of Agriculture, the Ministry of Rural Development, the Ministry of Urban Development, and the Ministry of Panchayati Raj.

Other reasons militate in favour of using a single institutional framework to address all development issues. Indeed, water is not the only issue that has the potential to go beyond the boundaries of a given panchayat, district or state. The same is true for other environmental resources such as biodiversity and

[126] eg R Watts, 'Contemporary Views on Federalism', in B de Villiers (ed), *Evaluating Federal Systems* (Dordrecht: Martinus Nijhoff, 1994) 1.

forests. There has been a trend in recent years to set up specific committees or institutions to address specific resources. This has an obvious logic but needs to be considered in the broader scheme of democratic governance. While there must be an obligation on panchayats sharing a watershed, a forest, a biodiversity reserve, to collaborate and cooperate on conservation and use issues, the basic unit of relevance remains the panchayat. If this is not done, the institutional framework will soon be made up of a series of uncoordinated and unrelated bodies that have overlapping mandates, for instance, because it is not possible to consider either biodiversity or forest without integrating water. In the case of biodiversity, this problem has been partly recognized. The biodiversity management committees are set up under panchayats or municipalities. The biodiversity regime also includes reservation for women and SCs/STs.[127] Yet, these committees are not committees of the panchayat and in fact they have restricted powers since their main function is only to prepare a biodiversity register.[128] Further, since they fall outside of the panchayat framework, they do not provide an appropriate forum for addressing cross-sectoral issues. In the case of forests, the programme known as joint forest management was initiated to involve village communities in the regeneration of degraded forests.[129] The joint forest management committees set up under this scheme are in principle registered as societies.[130] These societies are premised on being separate from the panchayat system. In fact, clarifications of the guidelines for joint forest management specifically make the case for this separation because joint forest management committees have a 'unique and separate non-political identity'.[131] The central problem is, however, that these committees do not have a mandate that allows them to take a broader view of forests within the context of other resources found in a given panchayat. This is all the more interesting because the same document makes the case for using the administrative and financial position and organizational capacity of the panchayats for the management of forest resources.[132]

The proposed reliance on panchayats does not imply that their powers should be absolute. Panchayats, like the state or Union governments are bound by environmental law. Thus, they are not in a position to unsustainably mine water. One way to establish this is by instituting a reserve at the level of decentralized bodies just like a reserve should be established at a national level. This would ensure that no democratically elected body, whether at the local or national level

[127] Biological Diversity Rules 2004, s 22(2).

[128] Biological Diversity Rules 2004, s 22(6).

[129] Ministry of Environment and Forests, Guidelines on Joint Forest Management, No. 6-21/89-P.P, 1 June 1990.

[130] Ministry of Environment and Forests, Guidelines on Joint Forest Management, No. 22-8/2000-JFM (FPD), 21 February 2000.

[131] Ministry of Environment and Forests, Guidelines Concerning the Strengthening of the Joint Forest Management Programme, 24 December 2002.

[132] ibid.

can take decisions that may be good for economic development in the short term but threaten the viability of development outcomes in the long term and are environmentally unsustainable.

Panchayats have also been repeatedly criticized in a number of states for failing to deliver development benefits more effectively than the state. This can be addressed over time by strengthening the institutions that have been put in place over the past 15 years in a process that will broadly contribute to better governance from the local to the national level. Additionally, consideration should be given to attributing powers directly to gram sabhas. This could be conceived as the fourth tier of decentralization that takes governance closer to people and removes some of the power that local elites tend to still wield in the panchayat system in a number of cases.[133] In other words, the constitutionally recognized gram sabha could be given much more specific and extensive powers than the legal framework provides today.[134] This proposal is different from ongoing calls for bypassing panchayats by setting up user groups.

E. The Need for a Comprehensive Water Law Framework

A series of broad bases for water law have been highlighted in the previous section. These would ensure that water regulation is conceived in ways which take into account the human rights, social, environmental, and equity aspects of water as well as its more traditional focus on the economic development contribution of the water sector.

This broad framework needs to be inscribed in contexts that ensure the application of basic principles throughout the water sector to all water, regardless of its source or use.[135] Differences can be made at the more specific implementation level depending on local circumstances but the basic framework should be the same because of the unitary nature of water and the water cycle.[136]

Current reforms have followed a process which could have led to the development of such a framework but that is not what has happened. The development of water policies at the Union and state levels has been a useful starting point in this process. Such efforts should, however, be the beginning of a process leading to the development of a framework water legislation rather than an end in itself. The difficulties associated with a process that stops mid-

[133] AK Vaddiraju & S Mehrotra, 'Making Panchayats Accountable' (2004) 39/37 *Economic & Political Weekly* 4139.

[134] For the case of Madhya Pradesh, eg A Behar, 'Gram Swaraj—Experiment in Direct Democracy' (2001) 36/10 *Economic & Political Weekly* 819.

[135] eg Mexico, Ley de aguas nacionales 1992, art 2.

[136] This was already acknowledged in Agenda 21, Report of the United Nations Conference on Environment and Development, United Nations, Rio de Janeiro, 3–14 June 1992, UN Doc. A/CONF.151/26/Rev.1 (Vol. 1), Annex II, para 18(35).

way are illustrated in the case of the MWRRA Act that proposes as operative principles the principles of the state water policy.[137]

In terms of water law development, the adoption of issue-specific acts must be stopped. This goes against the advice of the World Bank that expressly called for a 'cherry-picking' approach to water law reforms.[138] Nevertheless, the introduction of a comprehensive water law framework is necessary to remedy the shortcomings of the existing—and old—framework and provide answers to the new challenges that have surfaced in the past few decades. The first need is thus for a framework act that lays out the conceptual bases of water law, identifies all main operative principles and gives substance to the most important additions of the past few decades such as the recognition of the human right to water and the public trust nature of water in ways that take them beyond a general recognition that makes no real difference on the ground.

The various links from the local to the national level indicate that there is a need for a framework at the national level, which is the only level at which national level planning and coordination can be effectively undertaken. The need for some form of intervention of the Union government has been a recurrent feature of water policy over the past few decades.[139] These include the decision to step in and take a more pro-active role on drinking water in the early 1970s with the setting up of the Accelerated Rural Water Supply Programme, the adoption of the Water Act 1974 and the adoption—and revision—of a National Water Policy. At present, water remains a sensitive topic in political terms and the situation is not (yet) serious enough to ensure that the Union government musters the courage to introduce an all-India water legislation by following the same procedure adopted in 1974. Another possibility is the listing of water on the concurrent list in the same way that forests were listed.[140] This was specifically rejected by the National Commission for Integrated Water Resource Development Plan but is an issue which needs to be considered again in view of the new challenges that have developed in the water sector in the past decade.[141]

While Union legislation may be appropriate, it is not likely to be taken up in the short term. An alternative is thus the adoption by each state of a framework water legislation. In view of the current constitutional position where states have the main mandate to regulate water, this should in fact be the first obligation of states in the development of their water law. No further sectoral and issue-specific water-related legislation should be adopted until a framework legislation has been put in place. Such legislation must by definition be in consonance with existing

[137] ch 3.C.3, p. 99.

[138] World Bank, India—Water Resources Management Sector Review—Inter-sectoral Water Allocation, Planning and Management (Report No. 18322, 1998).

[139] R Iyer, *Towards Water Wisdom: Limits, Justice, Harmony* (New Delhi: Sage Publications, 2007) 176.

[140] Constitution (Forty-Second Amendment) Act, 1976.

[141] National Commission for Integrated Water Resource Development Plan, Report (New Delhi: Ministry of Water Resources, 1999) x.

constitutional principles and Supreme Court case law. In other words, even in the current dispensation where water is a state subject, a fair degree of uniformity is expected in state water legislation because most of the main principles have been defined at the national level in direct or indirect ways. The former include the specific recognition of the human right to water or the status of public trust while the latter include, for instance, principles developed in environmental law such as the prevention and precautionary principles.

The need to recognize the links from the local to the national level apply also beyond the national level. Indeed, the political challenge is similar within the country and between countries. Water has always been so important that it is politically extremely sensitive. This explains in part why international freshwater law has on the whole failed to move beyond issues that are directly related to cooperation on water which is in essence transboundary. Different reasons militate for a fundamentally different approach. Firstly, the distinction between transboundary watercourses and water found under national sovereignty is as artificial as the distinction between surface and groundwater. Secondly, the existence of a water cycle which is by definition global in scale calls for a completely different conceptualization of water in international law. This was always necessary but has acquired a new sense of urgency with some of the consequences of global warming being significant impacts on the global water cycle that have the potential to affect every country.

International water law needs to grow beyond its concerns for transboundary waters. The slow evolution that made international water law move beyond navigational issues to non-navigational aspects of transboundary waters needs to be quickly supplemented through a paradigm change. Indeed, unless international water law is able to address the most basic challenges that concern all states, like access to sufficient clean water for everyone and livelihood issues, it will lose its relevance in years to come. The need is not only for international water law to address water-specific issues that it has not addressed but also to address the cross-sectoral aspects of water law. This includes various links with international law concerning, for instance, human rights, the environment, health or agriculture.[142] This requires a definite effort to strengthen and broaden the scope of international water law beyond its narrow confines to ensure, for instance, that an international framework water convention can eventually be adopted.

The need for a comprehensive water law includes not only national and international water instruments that give water law a broader scope but also the inclusion of issues, which are directly related. This includes, for instance, impact assessment. Environmental and social impact assessment is an instrument that has been developed in the context of environmental law. Yet, it also applies to all

[142] Some of these concerns have been recognised in the Berlin Rules on Water Resources, International Law Association, Report of the Seventy-first Conference (2004) but this needs to be translated in binding instruments.

water projects that fall under the scope of the Environmental Impact Assessment Notification 2006. In other words, even where water law does not specifically address the environmental consequences of new water infrastructure, this is addressed under other frameworks and needs to be taken into account.

All the above remarks presuppose that a coordinated effort to take a comprehensive view of water law is adopted. This will have to happen at some point in the future but may well not take place immediately. In the meantime, it is at least imperative that any further sectoral water law reforms such as new laws addressing specific issues—irrigation, groundwater, drinking water—should address all the issues of that sector. This in itself will make a big difference to outcomes in conceptual and practical terms. Thus, in the case of irrigation, it is imperative to adopt reforms that take a broad view of the irrigation sector. This includes moving beyond laws with a limited focus like water user association laws that address a limited number of issues without taking into account the broader situation. This is unwelcome in practice and in law since it affects existing irrigation acts without specifically acknowledging all the consequences.

Finally, a broader legal framework needs to be inscribed in a broader institutional context. The existing framework is inappropriate insofar as, at the Union level, the main ministry concerned with water is the Ministry of Water Resources. Its focus on water as an input for economic development activities such as irrigation or hydropower makes it an inappropriate place for considering water in its various dimensions in its present form. Reforms must thus be introduced. Some efforts have already been made with the setting up of the MWRRA. Reforms must, however, go beyond this kind of intervention. Indeed, despite being broadly conceived insofar as regulatory tasks are concerned, the MWRRA is not a model for an institution overseeing water in its different dimensions. A broadly conceived institutional framework is one whose central institution would focus first on the hierarchically most important water use, namely drinking water. In other words, the institutional framework that can address the challenges of the future must directly reflect the priorities of the legal framework. This may imply setting up an umbrella water ministry or simply ensuring that priorities are reordered.

7
Conclusion

Water law in India is at a critical juncture. It has been in need of reform for a number of years but some of the most sensitive decisions were sidelined as long as possible by the various state governments. The direct links between water and life, water and livelihood, as well as water and agricultural production overall ensured that few governments were willing to take measures that would disturb existing arrangements. A number of reforms were necessary from the time of independence since, in particular, irrigation laws adopted under colonization focused on water as an economic resource and did not promote socially equitable results.

With time, the number of reasons for reforming existing water laws and for introducing additional laws has grown exponentially. The introduction of the Green Revolution whose success depended crucially on the sufficient availability of water is one of a number of elements that should have called for immediate water-related measures alongside the promotion of the new water-guzzling varieties that brought immense relief in terms of aggregate food security for India as a country. Indeed, while the additional water provided made a major contribution to the production of basic food crops, it also provided similar opportunities for even more water-intensive crops like cotton and sugarcane. Besides agriculture-specific arguments for water law reform, the progressive realization of the nexus between water and the environment did lead to some early measures in the context of the adoption of the Water Act 1974. This landmark decision did not, however, lead to the mainstreaming of environmental considerations into water law.

More generally, water has become an increasingly sensitive issue because of the multiple links between water and the realization of fundamental rights, water and the environment, water and food sovereignty, and water and economic development. The lack of a framework water law is thus increasingly evident and its absence has had some unfortunate consequences in recent years. One of the most damaging developments of the past two decades for water law has been the importance given to the Union and state water policies that have come to be regarded as some sort of a substitute for the missing framework legislation. This is inappropriate because it provides a new avenue for the introduction of a variety of measures that, in the worst case scenario, ignore Parliament, as has been the case in the context of drinking water. There is thus a problem of process in the adoption of measures that concern the majority of the population—rural dwellers—and focus on an issue of vital importance for everyone, the realization of the fundamental right to water. There is also a problem of substance where the adoption of measures that do not follow the normal law-making processes end up sidelining existing legal principles. Thus, the Swajaldhara Guidelines bemoan the fact that rural Indians consider water a 'social right' and fail to consider that it is not the 'public' that has affirmed that water is a fundamental right but the Supreme Court.

The necessity to update existing water laws, to introduce laws where they do not exist as in the case of drinking water and to introduce a framework water legislation at the Union or state levels is well established. The issue is not whether reforms are necessary or not but rather the kind of reforms that need to be introduced. In an immense country like India whose diversity extends to very diverse conditions from the point of view of water availability in different states, comprehensive debates need to take place to ensure that the laws adopted address the challenges faced by the nation and each individual state for decades to come. This is where ongoing water law reforms fail to match up to the ideal. Firstly, while water policy has been and is widely debated, this has by and large involved water sector experts. Broader debates that include everyone—since every single person has a direct and important stake in water—are yet to be taken up. This is where the non-involvement of Parliament in the context of drinking water reforms is unfortunate. Democratically elected MPs who have to answer to a constituency must be able to take the decisions that affect everyone's daily life. Secondly, where new water laws have been adopted by state legislative assemblies, the similarity in approach displayed by these laws is more than striking. Water user association legislation that looks alike throughout the country reflects two major shortcomings of the ongoing wave of reforms. On the one hand, these acts fail to build on each state's system of irrigation which has traditionally been different in different parts of the country. On the other hand, the adoption of a stream of water user association laws by various and diverse states while none of these states considers broader reforms of its irrigation legislation is symptomatic of an insufficiently developed legislative process. While it has been well established that existing irrigation acts fail to address the challenges of the twenty-first century, it is impossible to expect that the simple super-imposition of a set of institutional and management reforms will solve all existing problems and provide solutions for the additional challenges that the future will bring. The simple fact of not addressing irrigation legislation in its entirety ensures that the reform process is inappropriate at the outset and incapable of addressing the diversity of issues it is supposed to address.

One of the reasons for the problematic nature of ongoing water law reforms is that most of them are conceptually modelled on a restricted understanding of problems in the water sector. Thus the underlying issue is not water scarcity as conceived under water sector reforms but the need to understand water as a unitary substance from the most local uses to the global water cycle that no single nation can influence. This has significant implications with regard to the legal principles that are developed to address existing problems. Where the broader understanding of water is the basis of regulation, principles of non-appropriation like the principle of common heritage of humankind are more appropriate bases for legal measures than principles of appropriation linked to individual property rights.

Another problem linked to ongoing water law reforms is that they are based on the limited set of principles of water sector reforms. While water sector reforms provide insightful ways to address some of the problems of the water sector, they do not and cannot provide the basis for a broad-based revision of water law. This is due to the fact that water sector reforms mainly seek to address water sector problems from an economic perspective. As a result, water sector reforms are ill-suited to provide the basis for the implementation of fundamental human rights like the human right to water. Further, the set of principles they suggest are in large part economic principles and water legislation needs to be based on legal principles. While legal and economic principles inform each other and work to a large extent in tandem, one cannot serve as a replacement for the other.

The future is not necessarily bleak. Indeed, the very fact that water sector reforms have focused in recent years on the introduction of water laws has spurred a new realization of the importance of water law, an area of law relegated for too long to the shadows of other branches of law. The realization that water law is very important will likely lead to more broad-based debates in years to come. This is to be welcomed since water will be crucial to poverty eradication and the realization of the human right to water as well as the realization of a number of basic fundamental rights such as the human rights to life, food, and health.

It is, however, at this juncture that action needs to be taken to ensure that legal frameworks adopted in the next few years do not enshrine principles that will make it even more difficult than before to ensure the realization of the human right to water, water conservation, and food sovereignty. Water must imperatively be conceived in the twenty-first century as it has been for the past couple of thousand years at least, namely as a substance which bears very little relation to 'natural resources'. Indeed, water is too fundamental to human life and life on earth to be regulated along the principles adopted for the conservation and use of natural resources such as coal or other sectors such as electricity. Consequently, water is as a matter of principle *not* an economic good. In other words, the presumptions of existing water law reforms need to be reversed in favour of a system whose starting point is water's nature as a fundamental human right and a substance fundamental to the survival of life on earth. It is only *within* this context that water should be treated as an economic resource for the uses of water that are specifically focused on economic growth from industrial uses to large-scale cash crops.[1]

[1] cf Dublin Statement on Water and Sustainable Development, International Conference on Water and the Environment, Dublin, 31 January 1992, principle 4 making the opposite proposition. See ch 1.B.3, pp 25–6.

Selected Bibliography

Abernethy, CL, 'Constructing New Institutions for Sharing Water' in BR Bruns, C Ringler & R Meinzen-Dick (eds), *Water Rights Reform: Lessons for Institutional Design* (Washington, DC: International Food Policy Research Institute, 2005) 55.

Ambasta, P, PS Vijay Shankar & M Shah, 'Two Years of NREGA—The Road Ahead' (2008) 43/8 *Economic & Political Weekly* 41.

Armeni, C, The Right to Water in Italy (Geneva: International Environmental Law Research Centre, IELRC Briefing Paper 1, 2008) <available at www.ielrc.org/content/f0801.pdf>.

Baden-Powell, BH, *A Manual of Jurisprudence for Forest Officers* (Calcutta: Superintendant of Government Printing, 1882).

Bajpai, P & L Bhandari, 'Ensuring Access to Water in Urban Households' (2001) 36 *Economic & Political Weekly* 3774.

Bakker, KJ, *An Uncooperative Commodity—Privatizing Water in England and Wales* (Oxford: Oxford University Press, 2003).

Bandyopadhyay, J & S Perveen, 'Interlinking of Rivers in India—Assessing the Justifications' (2004) 39 *Economic & Political Weekly* 5307.

Bauer, CJ, *Siren Song: Chilean Water Law as a Model for International Reform* (Washington: Resources for the Future, 2004).

Behar, A, 'Revitalising Panchayati Rajs—Role of NGOs' (1998) 33/16 *Economic & Political Weekly* 881.

Behar, A, 'Gram Swaraj—Experiment in Direct Democracy' (2001) 36/10 *Economic & Political Weekly* 819.

Bhaduri, A & A Kejriwal, 'Urban Water Supply: Reforming the Reformers' (2005) 40/53 *Economic & Political Weekly* 5543.

Bhatia, R, 'Water and Economic Growth' in J Briscoe & RPS Malik, *Handbook of Water Resources in India—Development, Management and Strategies* (New Delhi: The World Bank and Oxford University Press, 2007) 99.

Bhatia, R, 'Water and Energy Interactions' in J Briscoe & RPS Malik, *Handbook of Water Resources in India—Development, Management and Strategies* (New Delhi: The World Bank and Oxford University Press, 2007) 206.

Bianchi, A, 'Human Rights and the Magic of *Jus Cogens*' (2008) 19 *European J Intl L* 491.

Bijoy, CR, 'Kerala's Plachimada Struggle—A Narrative on Water and Governance Rights' (2006) 42 *Economic & Political Weekly* 4332.

Birnie, PW & AE Boyle, *International Law and the Environment* (Oxford: Oxford University Press, 2002).

Black, M with R Talbot, *Water—A Matter of Life and Health* (New Delhi: Oxford University Press, 2005).

Bluemel, EB, 'The Implications of Formulating a Human Right to Water' (2004) 31/4 *Ecology LQ* 957.

Boesen, J & PE Lauridsen, '(Fresh)water as a Human Rights and a Global Public Good' in EA Andersen and B Lindsnaes (eds), *Towards New Global Strategies: Public Goods and Human Rights* (Leiden: Martinus Nijhoff Publishers, 2007) 393.

Briscoe, J & RPS Malik, *India's Water Economy—Bracing for a Turbulent Future* (New Delhi: The World Bank and Oxford University Press, 2006).

Brown Weiss, E, L Boisson de Chazournes & N Bernasconi-Osterwalder (eds), *Fresh Water and International Economic Law* (Oxford: Oxford University Press, 2005).

Budds, J & G McGranahan, 'Are the Debates on Water Privatization Missing the Point? Experiences from Africa, Asia and Latin America' (2003) 15/2 *Environment & Urbanization* 87.

Caflish, L, 'Règles générales du droit des cours d'eau internationaux' (1989) 219 *Recueil des cours—Académie de droit international* 2.

Caponera, DA, *National and International Water Law and Administration—Selected Writings* (The Hague: Kluwer Law International, 2003).

Chandra, N, 'The Evolving Institution of Groundwater Markets: A Model of Socially Embedded Exchange' (2004) 1 *Indian Juridical Rev* 111.

Coelho, K, 'The Slow Road to the Private—A Case Study of Neo-Liberal Water Reforms in Chennai' in P Cullet, A Gowlland-Gualtieri, R Madhav & U Ramanathan (eds), *Water Law at the Crossroads—National and International Perspectives With Special Emphasis on India* (New Delhi: Cambridge University Press, 2009) 81.

Craven, M, 'Some Thoughts on the Emergent Right to Water' in E Riedel & P Rothen (eds), *The Human Right to Water* (Berlin: Berliner Wissenschafts-Verlag, 2006) 37.

Cullet, P, *Differential Treatment in International Environmental Law* (Aldershot: Ashgate, 2003).

Cullet, P, A Gowlland-Gualtieri, R Madhav & U Ramanathan (eds), *Water Law for the Twenty-first Century: National and International Aspects of Water Law Reforms in India* (Abingdon: Routledge, forthcoming 2009).

Das, B & G Pangare, 'Privatisation: In Chhattisgarh, a River Becomes Private Property' (2006) 41/7 *Economic & Political Weekly* 611.

Dharmadhikary, S, *Unravelling Bhakra—Assessing the Temple of Resurgent India* (Badwani: Manthan, 2005).

Döckel, JAM, 'The Possibility of Trade in Water Use Entitlements in South Africa under the National Water Act of 1998', in S Perret, S Farolfi & R Hassan (eds), *Water Governance for Sustainable Development* (London: Earthscan, 2006) 35.

Donnelly, J, *Universal Human Rights in Theory and Practice* (Ithaca: Cornell University Press, 2nd ed. 2003).

Drèze, J, 'Democracy and the Right to Food', in P Alston & M Robinson (eds), *Human Rights and Development—Towards Mutual Reinforcement* (Oxford: Oxford University Press, 2005) 45.

Dubash, NK, 'Ecologically and Socially Embedded Exchange 'Gujarat Model' of Water Markets' (2000) 35 *Economic & Political Weekly* 1376.

Dubash, NK, 'The Electricity-Groundwater Conundrum: Case for a Political Solution to a Political Problem' (2007) 47/52 *Economic & Political Weekly* 45.

Dubash, NK, 'Independent Regulatory Agencies: A Theoretical Review with Reference to Electricity and Water in India' (2008) 43/40 *Economic & Political Weekly* 43.

Dubreuil, C, Synthesis on the Right to Water—Fourth World Water Forum, Mexico (Marseille: World Water Council, 2006).

Faruqui, NI, 'Islam and Water Management: Overview and Principles' in NI Faruqui, AK Biswas & MJ Bino (eds), *Water Management in Islam* (Tokyo: United Nations University Press, 2001) 1.

Faurès, J-M, M Svendsen & H Turral, 'Reinventing Irrigation', in D Molden (ed), *Water for Food, Water for Life* (London: Earthscan, 2007) 383.

Fidler, DP, 'A Kinder, Gentler System of Capitulations? International Law, Structural Adjustment Policies, and the Standard of Liberal, Globalized Civilization' (2000) 35 *Texas Intl L J* 387.

Finger, M & J Allouche, *Water Privatization—Trans-National Corporations and the Re-Regulation of the Water Industry* (London: Spon Press, 2002).

Fitzmaurice, M, 'General Principles Governing the Cooperation Between States in Relation to Non-Navigational Uses of International Watercourses' (2005) 14 *Ybk Intl Environmental L 2003* 3.

Fitzmaurice, M, 'The Human Right to Water' (2007) 18 *Fordham Environmental L Rev* 537.

Fitzmaurice, M & G Loibl, 'Current State of Development in the Law of International Watercourses' in SP Subedi (ed), *International Watercourses Law for the 21st Century—The Case of the River Ganges Basin* (Aldershot: Ashgate, 2005) 19.

Franck, TM, *Fairness in International Law and Institutions* (Oxford: Clarendon, 1995).

Galanter, M, *Competing Equalities—Law and the Backward Classes in India* (Berkeley: UC Press, 1984).

Gleick, PH, 'The Human Right to Water' (1999) 1/5 *Water Policy* 487.

Govinda Rao, M & UA Vasanth Rao, 'Expanding the Resource Base of Panchayats—Augmenting Own Revenues' (2008) 43/4 *Economic & Political Weekly* 54.

Ground Water Management and Ownership—Report of the Expert Group (New Delhi: Government of India, Planning Commission, 2007).

Guillet, D, 'Water Management Reforms, Farmer-Managed Irrigation Systems, and Food Security—The Spanish Experience' in L Whiteford & S Whiteford (eds), *Globalization, Water, and Health—Resource Management in Times of Scarcity* (Santa Fe: School of American Research Press, 2005) 185.

Gulati, A, R Meinzen-Dick & KV Raju, *Institutional Reforms in Indian Irrigation* (New Delhi: Sage, 2005).

Hemson, D, 'Water for All: From Firm Promises to "New Realism"?' in D Hemson et al. (eds), *Poverty and Water—Explorations of the Reciprocal Relationship* (London: Zed Books, 2008) 13.

Higgins, R, *Problems and Process—International Law and How We Use It* (Oxford: Clarendon, 1994).

Hildering, A, *International Law, Sustainable Development and Water Management* (Delft: Eburon, 2004).

Hirway, I, 'Ensuring Drinking Water to All: A Study in Gujarat' in KV Raju (ed), *Elixir of Life—The Socio-Ecological Governance of Drinking Water* (Bangalore: Books for Change, 2007) 74.

Hirway, I & S Goswami, 'Functioning of the Drinking Water Component of the Narmada Pipeline Project in Gujarat' (2008) 43/9 *Economic & Political Weekly* 51.

Hodgson, S, Legislation on Water Users' Organizations—A Comparative Analysis (Rome: FAO, FAO Legislative Study 79, 2003).

Hodgson, S, Land and Water—The Rights Interface (Rome: FAO, FAO Legislative Study 84, 2004).

Hodgson, S, Modern Water Rights—Theory and Practice (Rome: FAO, FAO Legislative Study 92, 2006).

Hoering, U & AK Schneider, King Customer? The World Bank's New Water Policy and its Implementation in India and Sri Lanka (Stuttgart: Brot für die Welt, 2004).

Holmes, S and CR Sunstein, *The Cost of Rights—Why Liberty Depends on Taxes* (New York: WW Norton, 1999).

Hooja, R, 'Below The Third Tier: Water Users Associations and Participatory Irrigation Management in India' (2004) 1 *Indian J Federal Studies*, available at http://www.jamiahamdard.edu/cfs/jour4-1_4.htm.

Ingram, H, JM Whiteley & R Perry, 'The Importance of Equity and the Limits of Efficiency in Water Resources', in JM Whiteley, H Ingram & R Perry eds, *Water, Place, and Equity* (Cambridge, Mass: MIT Press, 2008) 1.

Iyer, R, *Towards Water Wisdom: Limits, Justice, Harmony* (New Delhi: Sage Publications, 2007).

Iyer, R, 'Cauvery Award—Some Questions and Answers' (2007) 42/8 *Economic & Political Weekly* 639.

Iyer, R, 'Water: A Critique of Three Concepts' (2008) 48/1 *Economic & Political Weekly* 15.

Iyer, R (ed), *Water and the Laws in India* (New Delhi: Sage, forthcoming 2009)

Janakarajan, S & M Moench, 'Are Wells a Potential Threat to Farmers' Well-being? Case of Deteriorating Groundwater Irrigation in Tamil Nadu' (2006) 41/37 *Economic & Political Weekly* 3977.

Jha, N, 'Traditional Minor Irrigation Mechanisms: State Versus Community Conflicts' (2004) 1 *Indian Juridical Rev* 244.

Johnston, BR, 'The Commodification of Water and the Human Dimensions of Manufactured Scarcity' in L Whiteford & S Whiteford (eds), *Globalization, Water, and Health—Resource Management in Times of Scarcity* (Santa Fe: School of American Research Press, 2005) 133.

Johnstone, N & L Wood, 'Introduction' in N Johnstone & L Wood (eds), *Private Firms and Public Water—Realising Social and Environmental Objectives in Developing Countries* (Cheltenham: Edward Elgar, 2001) 1.

Kathpalia, GN & R Kapoor, Water Policy and Action Plan for India 2020: An Alternative (Delhi: Alternative Futures, 2002).

Keremane, GB & J McKay, 'Self-Created Rules and Conflict Management Processes: The Case of Water Users' Associations on Waghad Canal in Maharashtra, India' (2006) 22/4 *International Journal of Water Resources Development* 543.

Khera, R, 'Empowerment Guarantee Act' (2008) 43/35 *Economic & Political Weekly* 8.

Kiefer, T & C Brolmann, 'Beyond State Sovereignty: The Human Right to Water' (2005) 5/3 *Non-State Actors & Intl L* 183.

Koonan, S, 'Groundwater—Legal Aspects of the Plachimada Dispute' in P Cullet, A Gowlland-Gualtieri, R Madhav & U Ramanathan (eds), *Water Law at the Crossroads—*

National and International Perspectives With Special Emphasis on India (New Delhi: Cambridge University Press, 2009) 158.

Lang, A, 'The GATS and Regulatory Autonomy: A Case Study of Social Regulation of the Water Industry' (2004) 7 *J Intl Economic L* 801.

Malik, RPS, 'World Bank Policies and Lending Assistance' in J Briscoe & RPS Malik, *Handbook of Water Resources in India—Development, Management and Strategies* (New Delhi: The World Bank and Oxford University Press, 2007) 69.

McCaffrey, SC, 'The Human Right to Water' in E Brown Weiss, L Boisson de Chazournes & N Bernasconi-Osterwalder (eds), *Fresh Water and International Economic Law* (Oxford: Oxford University Press, 2005) 93.

McCaffrey, SC, *The Law of International Watercourses* (Oxford: Oxford University Press, 2007).

McDonald, DA & G Ruiters, 'Introduction: From Public to Private (to Public Again?)' in DA McDonald & G Ruiters (eds), *The Age of Commodity—Water Privatization in Southern Africa* (London: Earthscan, 2005) 1.

McDonald, DA & G Ruiters, 'Theorizing Water Privatization in Southern Africa' in DA McDonald & G Ruiters (eds), *The Age of Commodity—Water Privatization in Southern Africa* (London: Earthscan, 2005) 13.

McDonald, RSJ, 'Solidarity in the Practice and Discourse of Public International Law' (1996) 8 *Pace Intl L Rev* 259.

McKay, J & GB Keremane, 'Farmers' Perception on Self Created Water Management Rules in a Pioneer Scheme: The Mula Irrigation Scheme, India' (2006) 20 *Irrigation & Drainage Systems* 205.

Maltz, H, 'Porto Alegre's Water: Public and for All' in B Balanyá et al. (eds), *Reclaiming Public Water—Achievements, Struggles and Visions from Around the World* (Amsterdam: Transnational Institute and Corporate Europe Observatory, 2nd ed 2005) 29.

Mehta, L, 'Problems of Publicness and Access Rights: Perspectives from the Water Domain' in I Kaul et al. (eds), *Providing Global Public Goods—Managing Globalization* (Oxford: Oxford University Press, 2003) 556.

Meinzen-Dick, R & L Nkonya, 'Understanding Legal Pluralism in Water and Land Rights—Lessons from Africa and Asia' in B Van Koppen, M Giordano & J Butterworth (eds), *Community-Based Water Law and Water Resource Management Reform in Developing Countries* (Wallingford: CABI, 2007) 12.

Misra, S & B Goldar, 'Likely Impact of Reforming Water Supply and Sewerage Services in Delhi' (2008) 43/43 *Economic & Political Weekly* 57.

Moench, M, 'Approaches to Groundwater Management: To Control or Enable?' (1994) 29/39 *Economic & Political Weekly* A135.

Moench, M, 'Allocating the Common Heritage: Debates over Water Rights and Governance Structures in India' (1998) 33/26 *Economic & Political Weekly* A46.

Mohile, AD, 'Government Policies and Programmes' in J Briscoe & RPS Malik, *Handbook of Water Resources in India—Development, Management and Strategies* (New Delhi: The World Bank and Oxford University Press, 2007) 10.

Mosse, D, *The Rule of Water—Statecraft, Ecology and Collective Action in South India* (New Delhi: Oxford University Press, 2003).

Narain, V, *Institutions, Technology and Water Control* (New Delhi: Orient Longman, 2003).

National Commission for Integrated Water Resource Development Plan, Report (New Delhi: Ministry of Water Resources, 1999).

Narayanamoorthy, A & RS Deshpande, *Where Water Seeps!—Towards a New Phase in India's Irrigation Reforms* (New Delhi: Academic Foundation, 2005).

Olleta, A, 'The World Bank's Influence on Water Privatisation in Argentina—The Experience of the City of Buenos Aires' in P Cullet, A Gowlland-Gualtieri, R Madhav & U Ramanathan (eds), *Water Law at the Crossroads—National and International Perspectives With Special Emphasis on India* (New Delhi: Cambridge University Press, 2009) 230.

Orakhelashvili, A, *Peremptory Norms in International Law* (Oxford: Oxford University Press, 2006).

Pangare, V, N Kulkarni & G Pangare, *An Assessment of Water Sector Reforms in the Indian Context: The Case of the State of Maharashtra* (Geneva: UNRISD, 2004).

Pant, N, 'Impact of Irrigation Management Transfer in Maharashtra—An Assessment' (1999) 34/13 *Economic & Political Weekly* A-17.

Pant, N, 'Some Issues in Participatory Irrigation Management' (2008) 48/1 *Economic & Political Weekly* 30.

Paquin, M et al., Les accords sur l'investissement et les services et la gestion de l'eau dans les pays en développement—Défis et opportunités pour l'atteinte des Objectifs du Millénaire pour le développement en matière d'eau potable et d'assainissement (Cible 10) (Centre international Unisféra, 2004).

Pathak, S, 'Tehri Dam: Submersion of a Town, Not of an Idea' (2005) 40/33 *Economic & Political Weekly* 3637.

Petrella, R, *The Water Manifesto: Arguments for a World Water Contract* (London: Zed, 2001).

Pettiti, L-E & P Meyer-Bisch, 'Human Rights and Extreme Poverty', in J Symonides (ed), *Human Rights: New Dimensions and Challenges* (Aldershot: Ashgate, 1998) 157.

Phansalkar, S & V Kher, 'A Decade of the Maharashtra Groundwater Legislation' (2006) 2/1 *L Environment & Development J* 67, available at <http://www.lead-journal.org/content/06067.pdf>.

Planning Commission, Mid-Term Appraisal of the Tenth Five Year Plan (2005).

Planning Commission, *Towards Faster and More Inclusive Growth: An Approach to the Eleventh Five Year Plan 2007-2012* (New Delhi: Government of India, 2006).

Planning Commission—Government of India, *Eleventh Five Year Plan 2007–12—Volume I—Agriculture, Rural Development, Industry, Services and Physical Infrastructure* (New Delhi: Oxford University Press, 2008).

Planning Commission—Government of India, *Eleventh Five Year Plan 2007–12—Volume II—Social Sector* (New Delhi: Oxford University Press, 2008).

Prasad, K (ed), *Water Resources and Sustainable Development—Challenges for the 21st Century* (New Delhi: Shipra, 2003).

Prasad, N, 'Privatisation Results: Private Sector—Participation in Water Services After 15 Years' (2006) 24/6 *Development Policy Rev* 669.

Raju, KV, K Das & S Manasi, 'Emerging Trends in Rural Water Supply: A Comparative Analysis of Karnataka and Gujarat' in KV Raju (ed), *Elixir of Life—The Socio-Ecological Governance of Drinking Water* (Bangalore: Books for Change, 2007) 1.

Ramanathan, U, Legislating for Water: The Indian Context (Paper presented at the 3rd Common Property Conference, Washington, DC, 1992), available at <http://www.ielrc.org/content/w9201.pdf>.

Ramanathan, U, 'Displacement and the Law' (1996) 31/24 *Economic & Political Weekly* 1486.

Ramanathan, U, 'Demolition Drive' (2005) 40/27 *Economic & Political Weekly* 2908.

Ramanathan, U, 'A Word on Eminent Domain' in L Mehta (ed), *Displaced by Development – Confronting Marginalisation and Gender Injustice* (New Delhi: Sage, 2008) 133.

Randeria, S, 'Globalization of Law: Environmental Justice, World Bank, NGOs and the Cunning State in India' (2003) 51/3-4 *Current Sociology* 305.

Ratna Reddy, V & P Prudhvikar Reddy, 'How Participatory is Participatory Irrigation Management?—Water Users' Associations in Andhra Pradesh' (2005) 40/53 *Economic & Political Weekly* 5587.

Ray, I, '"Get the Price Right"—Water Prices and Irrigation Efficiency' (2005) 40/33 *Economic & Political Weekly* 3659.

Redgwell, C, *Intergenerational Trusts and Environmental Protection* (Manchester: Manchester University Press, 1999).

Riedel, E, 'The Human Right to Water and General Comment No. 15 of the CESCR' in E Riedel & P Rothen (eds), *The Human Right to Water* (Berlin: Berliner Wissenschafts-Verlag, 2006) 19.

Ruetschi, M, Déprivatisation de l'eau—L'expérience du Canton de Genève (Geneva: International Environmental Law Research Centre, IELRC Briefing Paper 2008–03, 2008) available at <http://www.ielrc.org/content/f0803.pdf>.

Salman, SMA & DD Bradlow, *Regulatory Frameworks for Water Resources Management—A Comparative Study* (Washington, DC: World Bank, 2006).

Salman, SMA & S McInerney-Lankford, *The Human Right to Water—Legal and Policy Dimensions* (Washington, DC: World Bank, 2004).

Sampat, P, '"Swa"-jal-dhara or "Pay"-jal-dhara—Sector Reform and the Right to Drinking Water in Rajasthan and Maharashtra' (2007) 3/2 *L Environment & Development J* 101, available at <http://www.lead-journal.org/content/07101.pdf>.

Sathe, SP, *Administrative Law* (New Delhi: Butterworths/Lexis-Nexis, 7th ed. 2004) 350.

Scanlon, J, A Cassar & N Nemes, Water as a Human Right? (Gland: IUCN, Environmental Policy and Law Paper No. 51, 2004).

Schlosberg, D, *Defining Environmental Justice—Theories, Movements, and Nature* (Oxford: Oxford University Press, 2007).

Scovazzi, T, The Concept of Common Heritage of Mankind and the Resources of the Seabed Beyond the Limits of National Jurisdiction (Paper prepared for the international workshop on Resources of the Seabed and Subsoil, Buenos Aires, 15–17 May 2006).

Sekhar, A, 'Development and Management Policies—Perspective of the Planning Commission' in J Briscoe & RPS Malik, *Handbook of Water Resources in India—Development, Management and Strategies* (New Delhi: The World Bank and Oxford University Press, 2007) 47.

Siddiqui, IA, 'History of Water Laws in India' in C Singh (ed), *Water Law in India* (New Delhi: Indian Law Institute, 1992) 289.

Shaban, A & RN Sharma, 'Water Consumption Patterns in Domestic Households in Major Cities' (2007) 42/23 *Economic & Political Weekly* 2190.

Shah, T, 'Institutional and Policy Reforms' in J Briscoe & RPS Malik, *Handbook of Water Resources in India—Development, Management and Strategies* (New Delhi: The World Bank and Oxford University Press, 2007) 306.

Shah, T, 'Groundwater Management and Ownership: Rejoinder' (2008) 48/17 *Economic & Political Weekly* 116.

Shah, T & B van Koppen, 'Is India Ripe for Integrated Water Resources Management? Fitting Water Policy to National Development Context' (2006) 41/30 *Economic & Political Weekly* 3413.

Singh, AK, *Privatization of Rivers in India* (Mumbai: Vikas Adhyayan Kendra, 2004).

Singh, C, *Water Rights and Principles of Water Resources Management* (Bombay: Tripathi, 1991).

Singh, C (ed), *Water Law in India* (New Delhi: Indian Law Institute, 1992).

Smets, H, 'Le droit à l'eau, un droit pour tous en Europe' (2007) 37/2-3 *Environmental Policy & L* 223.

Smith, JA and JM Green, 'Free Basic Water in Msunduzi, KwaZulu-Natal: Is it Making a Difference to the Lives of Low-income Households?' (2005) 7 *Water Policy* 443.

Sohnle, J, *Le droit international des ressources en eau douce: Solidarité contre souveraineté* (Paris: La Documentation française, 2002).

D'Souza, R, *Interstate Disputes over Krishna Waters—Law, Science and Imperialism* (New Delhi: Orient Longman, 2006).

Subedi, SP (ed), *International Watercourses Law for the 21st Century—The Case of the River Ganges Basin* (Aldershot: Ashgate, 2005).

Swatuk, LA, 'The New Water Architecture of SADC' in DA Mcdonald & G Ruiters (eds), *The Age of Commodity—Water Privatization in Southern Africa* (London: Earthscan, 2005) 43.

Takacs, D, 'The Public Trust Doctrine, Environmental Human Rights, and the Future of Private Property' (2008) 16 *New York University Environmental L J* 711.

Takacs, D, 'Environmental Aspects of Water Sector Reforms', in P Cullet, A Gowlland-Gualtieri, R Madhav & U Ramanathan (eds), *Water Law for the Twenty-first Century: National and International Aspects of Water Law Reforms in India* (Abingdon: Routledge, forthcoming 2009).

Tarlock, AD, 'Water Transfers: A Means to Achieve Sustainable Water Use' in E Brown Weiss, L Boisson de Chazournes & N Bernasconi-Osterwalder (eds), *Fresh Water and International Economic Law* (Oxford: Oxford University Press, 2005) 35.

Tiwari, R, 'Explanations in Resource Inequality—Exploring Schedule Caste Position in Water Access Structure' (2006) 2/1 *Intl J Rural Management* 85.

United Nations, *Water—A Shared Responsibility* (Paris: UNESCO, 2006).

United Nations, *Water for People—Water for Life* (Paris: UNESCO, 2003).

United Nations Development Programme, *Human Development Report 2006—Beyond Scarcity: Power, Poverty and the Global Water Crisis* (New York: UNDP, 2006).

United Nations Development Programme, *Human Development Report 2007/2008* (New York: UNDP, 2007).

United Nations Environment Programme, *Global Environment Outlook 3* (London: Earthscan, 2002).

United Nations Environment Programme, *Global Environment Outlook—GEO4—Environment for Development* (Nairobi: UNEP, 2007).

Upadhyay, V, 'Water Management and Village Groups: Role of Law', 37 *Economic and Political Weekly* 4907 (2002).

Upadhyay, V, 'Customary Rights over Tanks' (2003) 38/44 *Economic & Political Weekly* 4643.

Upadhyay, V, Law under Globalization—Assessing 'Donor Supported' Law Making and Judicial Behaviour in India (New Delhi: Social Watch Coalition, 2008).

Upadhyay, V, 'Canal Irrigation, Water User Associations and Law in India—Emerging Trends in Rights-based Perspective' in P Cullet, A Gowlland-Gualtieri, R Madhav & U Ramanathan (eds), *Water Law at the Crossroads—National and International Perspectives With Special Emphasis on India* (New Delhi: Cambridge University Press, 2009) 110.

Vaddiraju, AK & S Mehrotra, 'Making Panchayats Accountable' (2004) 39/37 *Economic & Political Weekly* 4139.

Vandenhole, W & T Wielders, 'Water as a Human Right—Water as an Essential Service—Does it Matter?' (2008) 26/3 *Netherlands Q Human Rights* 391.

Vani, MS, 'Reviving Customary Law: Enabling Law for Water Harvesting' in A Agarwal, S Narain and I Khurana (eds), *Making Water Everybody's Business—Practice and Policy of Water Harvesting* (New Delhi, Centre for Science and Environment, 2001) 332.

Varughese, GC, 'Water and Environmental Sustainability' in J Briscoe & RPS Malik, *Handbook of Water Resources in India—Development, Management and Strategies* (New Delhi: The World Bank and Oxford University Press, 2007) 184.

Vörösmarty, CJ, 'Fresh Water' in R Hassan, R Scholes & N Ash, *Ecosystems and Human Well-being: Current State and Trends—Millennium Ecosystem Assessment Series Volume 1* (Washington: Island Press, 2005) 170.

Ward, C, *Reflected in Water—A Crisis of Social Responsibility* (London: Cassell, 1996).

Watts, R, 'Contemporary Views on Federalism', in B de Villiers (ed), *Evaluating Federal Systems* (Dordrecht: Martinus Nijhoff, 1994) 1.

Whiteley, JM, H Ingram & R Perry (eds), *Water, Place, and Equity* (Cambridge, Mass: MIT Press, 2008).

Woodhouse, EJ, 'The "Guerra del Agua" and the Cochabamba Concession: Social Risk and Foreign Direct Investment in Public Infrastructure' (2003) 39/2 *Stanford J Intl L* 295.

World Bank, India—Water Resources Management Sector Review—Report on the Irrigation Sector (Report No. 18416 IN, 1998).

World Bank, India—Water Resources Management Sector Review—Urban Water Supply and Sanitation Report (Report No. 18321, 1998).

World Bank, India—Water Resources Management Sector Review—Inter-sectoral Water Allocation, Planning and Management (Report No. 18322, 1998).

World Bank, India—Water Resources Management Sector Review—Rural Water Supply and Sanitation Report (Report No. 18323, 1998).

World Bank, India—Water Resources Management Sector Review—Groundwater Regulation and Management Report (Report No. 18324-IN, 1998).

World Bank, India—Water Resources Management Sector Review—Initiating and Sustaining Water Sector Reforms (Report No. 18356-IN, 1998).

World Bank, Efficient, Sustainable Service for All? An OED Review of the World Bank's Assistance to Water Supply and Sanitation (Report No. 26443, 2003).

World Panel on Financing Water Infrastructure, Financing Water for All (Marseille: World Water Council, 2003).

Wouters, P, The Legal Response to International Water Scarcity and Water Conflicts—The UN Watercourses Convention and Beyond (University of Dundee: Water Law and Policy Programme, 2003).

Zérah, M-H, *Water—Unreliable Supply in Delhi* (New Delhi: Manohar, 2000).

NOTES

- A number of recent documents concerning water sector reforms and water law reforms in India and at the international level have been uploaded on the website of the Water Law Research Partnership funded by the Swiss National Science Fund. A complete list can be found at <http://www.ielrc.org/water/docs.htm>.
- Most of the recent legislation adopted by Indian states cited in this book can be found online at <http://www.ielrc.org/water/doc_states.php>.
- Most of the recent water documents adopted at the union level cited in this book can be found online at <http://www.ielrc.org/water/doc_goi.php>.
- A number of the Indian Supreme Court and high court cases related to water cited in this book can be found respectively at <http://www.ielrc.org/water/doc_sccases.php> and <http://www.ielrc.org/water/doc_hccases.php>.
- Reference to the 'eleventh plan' or to another of the plans should be understood as a reference to the 'eleventh five year plan'.
- At the end of 2008 the exchange rate for the Indian rupee was around 72 to the British pound and 50 the United States dollar.

Index